NONVOLATILE MEMORY DESIGN

MAGNETIC, RESISTIVE, AND PHASE CHANGE

NONVOLATILE MEMORY DESIGN

MAGNETIC, RESISTIVE, AND PHASE CHANGE

Hai Li • Yiran Chen

CRC Press
Taylor & Francis Group
Boca Raton London New York

CRC Press is an imprint of the
Taylor & Francis Group, an **informa** business

CRC Press
Taylor & Francis Group
6000 Broken Sound Parkway NW, Suite 300
Boca Raton, FL 33487-2742

First issued in paperback 2017

© 2012 by Taylor & Francis Group, LLC
CRC Press is an imprint of Taylor & Francis Group, an Informa business

No claim to original U.S. Government works
Version Date: 20111025

ISBN 13: 978-1-138-07663-1 (pbk)
ISBN 13: 978-1-4398-0745-3 (hbk)

Visit the Taylor & Francis Web site at
http://www.taylorandfrancis.com

and the CRC Press Web site at
http://www.crcpress.com

Contents

Preface

Among the numerous exciting inventions that dramatically changed our everyday life in the 20th century, semiconductor technology is one of the most influential achievements. In 1970, the first two important embodiments of semiconductor memory—1-kilobyte (KB) dynamic random access memory (DRAM) and 256-KB static random access memory (SRAM)—were invented by Intel/IBM and Fairchild, respectively. Because both DRAM and STRAM require power to maintain the stored information, they are categorized as *volatile memory*.

Another category of semiconductor memory is called *nonvolatile memory*, which can keep information stored even when power is disabled. One example of nonvolatile memory is *flash memory*, which is based on electrically erasable programmable read-only memory (EEPROM) technology. The first commercial EEPROM product was introduced by Intel in 1983, with a 16-KB capacity.

The capacity of semiconductor memory has achieved unprecedented increase: the latest flash memory capacities have reached 512 gigabytes (GB). This means that flash memory capacity has doubled every 12 months continuously for 25 years. The cost per bit of flash memory was also reduced by almost 200,000 times. Driven by the ever-exploding demand for higher capacity nonvolatile memory in pervasive computing and communication devices, flash memories have been the fastest growing segment in the global semiconductor industry: in 2007, the total capacity of flash memory shipped to market exceeded DRAM.

As complementary metal-oxide semiconductor (CMOS) integration technology enters the nano regime, engineers are facing technology challenges unlike any encountered in the 40-year history of the semiconductor industry: large process and device variation, high leakage power, complex manufacturing processes, etc. These challenges not only slow down the technology scaling (Moore's law) but also introduce significant reliability issues. For example, one fundamental limitation of flash memory is called *endurance*, which is measured by the number of times the flash memory cell can be programmed. When process technology scales down to 45 nm and beyond, the endurance of a multilevel flash memory cell is degraded to about 1,000 cycles. Very low endurance introduces high management costs for memory operation and severely limits the application of flash memory.

Another technical limitation of flash memory is the slow programming (writing) speed. Usually it takes up to 200 ms to program one flash memory cell. An intelligent control scheme like program–verify–reprogram ... is usually used to overcome the instability of flash memory cells due to process variation, etc. However, such a complex scheme may take up to 1 ms in order to successfully write one bit. Slow programming speed significantly limits the applications of flash memory, compared to its competitors, for example, DRAM, with a 15 ns access time. A natural solution is to introduce parallelism—programming multiple memory cells simultaneously—called *page mode* into the write operation of flash memory. Page mode improves memory bandwidth by sacrificing the capability to randomly access the individual bit.

Because traditional memory technologies—for example, DRAM, SRAM, and flash—have approached their limits, a new concept called *universal memory* is considered. Universal memory is believed to have characteristics such as high density (low cost), high speed (for both read and write operations), low power (both leakage and dynamic powers), random accessibility, nonvolatility, and unlimited endurance that can meet the requirements of various applications, from a large, expensive supercomputer to a low-cost, ubiquitous, consumer handheld device. Because none of the currently available memory technologies are promising candidates for universal memory, researchers are seeking alternative solutions.

The motivation of this book is to introduce those emerging memory technologies that are close to the definition of universal memory. At the time of this writing, some of these candidates have become commercially available, for example, phase change memory (PCM) and toggle-mode magnetic random access memory (TM-MRAM). Other candidates are still in the research stage but have demonstrated very promising characteristics and gained attention from both industry and academia, for example, spin-torque transfer RAM (STT-RAM) and resistive RAM (R-RAM). In order to become the winner in this $60 billion market (the total market capital of semiconductor memory in 2007), billions of dollars have been invested in this area by almost all major semiconductor and storage companies.

This book is intended to provide the following information to our readers:

- Background and fundamental mechanisms of emerging nonvolatile memory technologies; for example, physics, device structure, materials, and operating mechanisms.
- Manufacture and process technologies; for example, manufacturing processes for special storage devices and CMOS integration technology.
- Circuit design techniques; for example, memory array design, read/write circuitry, and testing strategies.
- Potential applications and future trends.

Rather than offering a review (or a simple collection) of the latest publications on emerging technologies, this book reflects the state-of-the-art experience of two authors—Dr. Hai Li and Dr. Yiran Chen—on the research on emerging nonvolatile memory technologies. Both Dr. Li and Dr. Chen have worked with the Alternative Technology Group at Seagate Technology where they serve as design lead and architect, respectively. Currently Dr. Li is an assistant professor in the Department of Electrical and Computer Engineering,

Polytechnic Institute of New York University. Dr. Chen is an assistant professor in the Department of Electrical and Computer Engineering, University of Pittsburgh. They have authored or coauthored more than 70 U.S. patents in the emerging nonvolatile memory area, more than 80 refereed technical papers, and three book chapters. Their works have received two best paper awards and three best paper nominations.

This book assumes a basic knowledge of semiconductor device physics and metal-oxide semiconductor field-effect transistor (MOSFET), though we briefly introduce this information in Chapter 1. We target all semiconductor memory designers who are willing to know the latest progress and predictions in emerging nonvolatile memory technologies or the nonvolatile memory designers who want to stay current regarding the technology trends in their working area. This book will also be a good reference for undergraduate/graduate students who are studying or doing research on nonvolatile memory technologies.

The book is organized into six chapters, mainly based on different memory types. The necessary knowledge of semiconductor device physics and MOSFET operation is included in Chapter 1. Chapters 2–4 cover PCM, STT-RAM, and R-RAM, respectively. Chapter 5 introduces an important mutant of R-RAM—the *memresistor*—which is actually an analog memory device whose state is determined by its continuous historical status. In Chapter 6, we provide an assessment as to who we believe will be the winner in this race.

Author Biographies

Hai Li received a B.S. degree in electrical engineering and an M.S. degree in microelectronics from Tsinghua University, Beijing, China. She received a Ph.D. degree from the Electrical and Computer Engineering Department of Purdue University, West Lafayette, Indiana.

Dr. Li was with Qualcomm CDMA Technologies (QCT) of Qualcomm Inc., San Diego, California, where she mainly worked on memory design and design methodology at 65-nm technology node and beyond. Then she joined the Microprocessor Product Group (MPG) of Intel Corp., Santa Clara, California, where she contributed to the success of Centrino Core Duo, the first 45-nm central processing unit in the world. She then joined the Alternative Technology Group of Seagate Technology LLC, Bloomington, Minnesota, where she worked as the circuit design lead for a next-generation nonvolatile memory technology development team.

Dr. Li became an assistant professor in 2009 in the Department of Electrical and Computer Engineering at Polytechnic Institute of New York University. Her research interests include architecture/circuit/device co-optimization, emerging memory design, 3D integration technology and design, and design for new devices.

Dr. Li has published more than 50 technical papers in refereed journals and conferences, 49 U.S. patents (25 granted), and one Seagate Trade Secret. She also authored two book chapters. Dr. Li received two best paper awards and three best paper awards from the International Symposium on Quality Electronic Design (ISQED), the International Symposium on Low Power Electronics and Design (ISLPED), Design, Automation & Test in Europe (DATE), and the Asia and South Pacific Design Automation Conference (ASPDAC). Dr. Li has been on technical program committees of more than 16 international conference series.

Yiran Chen received B.S. and M.S. degrees in Electronics Engineering from Tsinghua University, Beijing, China, and a Ph.D. degree from the Electrical and Computer Engineering Department of Purdue University, West Lafayette, Indiana.

He was with the PrimeTime group of Synopsys Inc., Sunnyvale, California, where he developed the market-dominating static timing analysis tool PrimeTime and the award-winning statistical static timing analysis tool PrimeTimeVX. He then joined the alterative technology group of Seagate Technology LLC, Bloomington, Minnesota, where he worked on the next-generation nonvolatile memory technology as a memory architect.

Currently, Dr. Chen is an assistant professor in the Electrical and Computer Engineering Department at the University of Pittsburgh. His research interests include very-large-scale integration design/computer-aided design for nanoscale technologies, low-power circuit design and architecture, emerging memory technologies, and nanoscale reconfigurable computing systems and sensor systems.

Dr. Chen has published more than 70 technical papers in refereed journals and conferences, edited or coauthored 1 book and 3 book chapters, has 63 U.S. and international patents (33 granted), and one Seagate Trade Secret. He has been on the technical program committees of more than 13 international conference series and the editor of the *Journal of Convergence Information Technology.*

Dr. Chen received the Hot 100 Products of 2006—PrimeTimeVX Award from Electronic Design News (EDN) and the was a finalist for the prestigious 2007 DesignVision Award from the International Engineering Consortium (IEC). He also received the PrimeTimeVX—EDN 100 Hot Products Distinction from Synopsys Inc. He received two best paper awards and four best paper nominations from the International Symposium on Quality Electronic Design (ISQED), Design, Automation & Test in Europe (DATE), the Asia and South Pacific Design Automation Conference (ASPDAC), and the International Symposium on Low Power Electronics and Design (ISLPED). His invention of the spintronic memristor was reported by *IEEE Spectrum* in March 2009.

1

Introduction to Semiconductor Memories

1.1 Classification and Characterization of Semiconductor Memories

Semiconductor memories that can be used to store user data, program code, and other information play an important role in computing systems and embedded systems. A memory system can store millions of words, and each word contains many bits. Because of the rapid development of information technology, the requirements for speed and capacity of the memory system are increasing. Therefore, memory manufacturers and designers need to continuously improve the process technologies and design methodologies in order to enable technology scaling and provide better performance by paying a lower cost.

Based on the functionalities, semiconductor memory can be divided into read-only memory (ROM) and read-write memory, as shown in Figure 1.1. Read/write memory is sometimes called *random-access memory* (RAM), which is a little confusing because ROM can also be randomly accessed.

The core of a semiconductor memory system is its memory array, shown in Figure 1.2. Here, each storage unit in the array is called a *cell*. One memory cell can usually store one bit of data. We call this a *single-level cell* (SLC). Some memory technologies can store more than one bit of data in one memory cell, and this is called a *multilevel cell* (MLC). A memory array is commonly implemented in the form of square in order to reduce the complexity of the decoder circuitry and to reduce the maximum interconnecting wire length.

In addition to the memory array, address decoders are necessary in order to identify the memory cells to be accessed. Write circuitry and sense amplifiers are respectively used to program the memory cells and read out the stored data. Interface circuits are required to connect the array to external signals. Other circuits, such as control logic, are also required in order to make the memory system function properly.

Currently, the most commonly used semiconductor memories are ROM (including some programmable ROM), static RAM (SRAM), dynamic RAM (DRAM), and nonvolatile memories.

1.1.1 Read-Only Memory

Read-only memory can be simply implemented by using only one transistor per bit as shown in Figure 1.3. Therefore, it can reach very high density

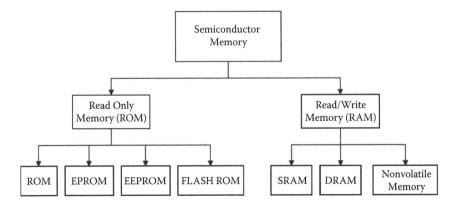

FIGURE 1.1
Categorization of semiconductor memory.

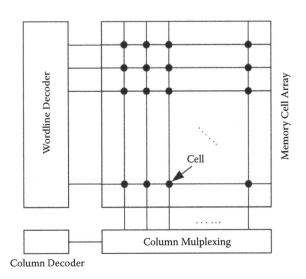

FIGURE 1.2
Basic memory structure.

compared to other memory structures. In Figure 1.3, the 0/1 of each bit is realized through the absence/presence of an N-channel metal–oxide–semi-conductor field-effect transistor (NMOS) transistor in the corresponding position. When a word line is activated, the absence/presence of an NMOS transistor causes the corresponding isolated/connected bit line to ground. Hence, the different voltages can be generated at bit lines, which are trans-lated to different binary values after amplification by the inverters connected to the bit lines. For example, the content of the ROM in Figure 1.3 is

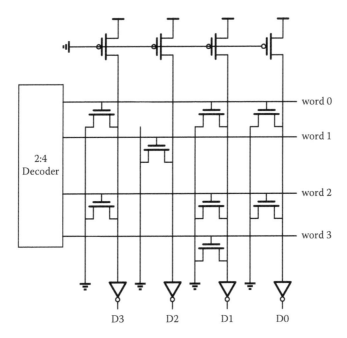

FIGURE 1.3
Simple ROM structure.

```
Word0:1011
Word1:0100
Word2:1011
Word3:0010
```

Instead of utilizing the absence/presence of transistors to achieve logic 0/1, ROM can also be implemented by the absence/presence of the contact between the metal layer and the gate of each transistor. In other words, a transistor exists in every cell position of the array, but not all of them contact the metal layer and make the connection. This design has a uniform structure and can provide more flexibility: if the content of the ROM needs to be changed, only the contact layer needs to be replaced, without impacting any other layers.

ROM can be used in applications in which the data will be kept permanently unchanged; for example, the lookup table content for some decoders or the program code for a vending machine. The data stored in ROM are nonvolatile, which means that even if we remove the power supply of the memory system, the data inside will not be lost.

The data in ROM are predetermined at the design stage and cannot be changed after fabrication. However, many applications require reprogrammability. For instance, the secret key stored in a smartcard should be nonvolatile. However, each smartcard user needs to own a unique key, which can only be determined postfabrication. In such a situation, ROM is not enough and reprogrammability is essential.

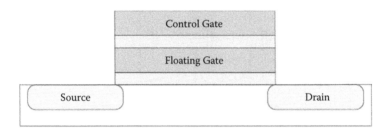

FIGURE 1.4
Floating gate structure of EPROM.

1.1.2 Erasable Programmable ROM and Flash Memory

Erasable programmable ROM (EPROM) is nonvolatile as well as programmable. It uses a special storage cell in order to achieve certain useful characteristics as shown in Figure 1.4. This cell is made of a transistor with two gates. On the top there is a primary gate (or control gate), and an insulated gate, called a *floating gate*, is placed in the middle. A binary value of an EPROM depends on the threshold voltage of the transistor, which is determined by whether the floating gating charged or not. Because the floating gate is insulated, the electrons within it can be kept from escaping for a long time; that is, more than 10 years.

When programming an EPROM cell, we can apply a high voltage (approximately 25 V) to the control gate. The electrons in the substrate can jump into the floating gate because of a process called *avalanche injection*, where the electron-hole pairs are generated under high voltage. As a result, the threshold voltage of the transistor can be increased to a level to keep the transistor in "off" mode.

To erase the data, the EPROM has to be exposed to a strong ultraviolet (UV) light from a mercury vapor light source for a certain time until the electrons in the floating gate gain enough energy to escape from it. Accordingly, the chip package of EPROM usually has a small quartz window to admit UV light for erasure. The erase operation can take 20–30 minutes.

This type of EPROM uses the presence/absence of electrons in the floating gate to represent the binary data. One major disadvantage is that the EPROM chip has to be taken out of the system and exposed to UV light when erasure is needed.

1.1.3 Electrically Erasable Programmable ROM

In many applications, it is not realistic to remove the memory chip from the system to erase the stored data. The advent of electrically erasable programmable ROM (EEPROM), which can be programmed and erased within the system by electrical methods, has solved this problem.

The basic storage cell in EEPROM is similar to that in EPROM, as shown in Figure 1.5. There are two gates: the control gate on the top is used for programming/erasing and the floating gate in the middle is for data storage.

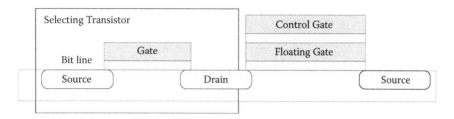

FIGURE 1.5
Cell structure of EEPROM.

However, the oxide layer between the floating gate and the transistor channel is much thinner than that of an EPROM, which makes electron transportation easier.

During a programming process, a high-voltage pulse is applied to the control gate and the transistor drain is connected to the ground. As a result, the electrons flow from the drain to the floating gate. Conversely, when connecting the control gate to the ground and applying the drain of the transistor to a high-voltage pulse, the electrons flow from the floating gate to the drain and, hence, the stored data is erased. Here, both programming and erasing operations can be completed by an electrical approach instead of by exposure to the UV light as EPROM. The high-voltage pulse is usually generated inside the chip with the aid of a charge pump circuit.

In some designs, a selecting transistor is used in each cell to prevent unnecessary disturbance among different memory cells. In such a situation, the cell size of EEPROM is larger than that of EPROM.

Flash memory, which is widely used in memory cards, universal serial bus (USB) storage drives, and MP3 players, is a specific kind of EEPROM. Instead of erasing byte by byte as EEPROM does, flash memory can erase the data in larger blocks simultaneously. Because the duration of erasing cycle for EEPROM bits is long, block-based erasing gives flash memory a great advantage in speed when a large amount of data needs to be written. Although flash memory may not be used in applications requiring byte-level random access (i.e., programming code storage), it still offers a high-capacity nonvolatile storage solution at low cost and with good access performance.

1.1.4 Static RAM

Frequent and high-speed read/write operations are extremely important in many applications, such as caches in high-performance processors. In such situations, static RAM is widely used. The word *static* here means that the data can self-maintain in SRAM as long as power is supplied. No refresh is needed in SRAM design.

Figure 1.6 shows the cell structure of an SRAM cell. Two inverters (M1, M2 and M3, M4) are cross-coupled to form a feedback loop for the purpose of storing the binary data. Another two NMOS transistors (M5 and M6) are

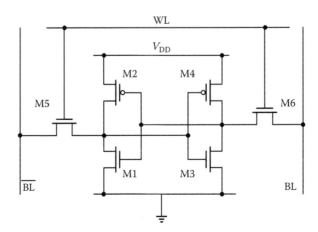

FIGURE 1.6
SRAM cell structure.

used as selecting switches to control access to the cell. Each SRAM cell has two bit lines, namely, BL and $\overline{BL}$.

When the word line (WL) is connected to the ground and the selecting transistors M5 and M6 are turned off, the SRAM cell is in a standby state. As long as the power supply V_{DD} is provided, the cross-coupled inverters will reinforce each other. Therefore, the binary data Q and $\overline{Q}$ remain at their original values.

For a read operation, both bit lines (BL and $\overline{BL}$) need to be precharged, for example, both bit lines (BL and BL) need to be precharged to a certain voltage (i.e., V_{DD}), before the start of the reading cycle. A read operation will start with connecting the WL to V_{DD} so that both selecting transistors M5 and M6 are turned on. Here we assume that the memory cell stores '1'. Due to the cross-coupled structure, the voltages at Q and $\overline{Q}$ are opposite and we have Q = 1 and $\overline{Q}$ = 0. The difference between Q and $\overline{Q}$ will affect the bit lines; that is, BL remains the precharged high voltage and $\overline{BL}$ is discharged toward 0. BL and $\overline{BL}$ are sent to the inputs of a sense amplifier to extract the binary value stored in the SRAM cell. Note that in this example, when $\overline{BL}$ is pulled down by the transistors M5 and M1, the voltage at node $\overline{Q}$ tends to rise because $\overline{BL}$ is precharged to a high voltage. Therefore, in SRAM design, M1 must be stronger than M5 to prevent the inverter composed by M3 and M4 from flipping over. To be more specific, the size ratio of M1 and M5 must guarantee that the voltage at node $\overline{Q}$ in the scenario of the given example stays below the switching threshold of the inverter M3/M4 during the read operation.

In a write operation, the value to be written should be put on the bit lines first. Assume that the data previously stored in the SRAM cell is '0', which means Q = 0 and $\overline{Q}$ = 1. Now we want to write '1' to the same SRAM cell and hence set BL to '1' and $\overline{BL}$ to '0'. Once the WL is connected to a high voltage and the transistors M5 and M6 are turned on, the new values on the bit lines need to override the original data stored in the feedback loop. Because Q cannot be pulled up to V_{DD} by BL through M3 and M6, the only way to flip the data is to pull $\overline{Q}$ down by $\overline{BL}$ through M5 and M4. This requires M4 to be much weaker than M5. Once $\overline{Q}$ is lower than the switching threshold of the inverter M3/M4, M3 is turned off and M4 is turned on, which helps to increase the voltage at Q.

The feedback loop in an SRAM cell needs to be powered on to maintain its state. The data will be lost if the power is turned off. Hence, we say that SRAM is a volatile memory device.

Because there are six transistors in one SRAM cell, it has less density and is more expensive compared to the other memory technologies. However, SRAM can achieve the best read/write speed and is still widely used in applications where high performance is the most critical factor.

1.1.5 Dynamic RAM

Among all of the semiconductor random access read/write memories, the storage density of DRAM could be the highest. Each DRAM cell is composed of only one transistor and one capacitor. Unlike SRAM, which utilizes the cross-coupled inverters to store data, the content in a DRAM cell is stored as the charge on the capacitor. However, the charge on the capacitor tends to leak away and, hence, the DRAM memory cell needs to be refreshed periodically. This is why it is called dynamic RAM.

The cell structure of a one-transistor DRAM is shown in Figure 1.7. Similar to SRAM, the gate of the transistor is connected to the word line. The transistor behaves as a selecting switch to control access to the cell. One diffusion area of the transistor is connected to the bit line and the other is connected to the capacitor.

The write operation in DRAM is relatively simple. Based on the data to be stored, we set the bit line to high or low voltage. After turning on the selecting transistor, the corresponding capacitor in the same cell is charged to a

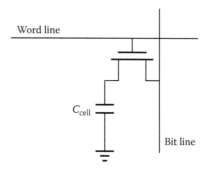

FIGURE 1.7
DRAM cell structure.

high voltage or discharged to the ground. The previous value is overwritten by the new value.

In a read operation, the bit line is first precharged to $V_{DD}/2$, and then the word line is connected to a high voltage. This turns on the selecting transistor, and the capacitor and the bit line will share the charges. Depending on the voltage that the capacitor was previously charged to, that is, V_{DD} or 0, eventually the voltage on the bit line becomes $V_{DD}/2 \pm \Delta V$. Here ΔV can be calculated by:

$$\Delta V = \frac{V_{DD}}{2} \frac{C_{cell}}{C_{cell} + C_{bitline}}$$

where C_{cell} is the capacitance of the capacitor and $C_{bitline}$ is the capacitance of the bit line. A sense amplifier is needed to amplify this voltage difference. After each read operation, the charge on the capacitor is influenced by the bit line, so the data need to be written back to the cell by charging/discharging the capacitor to its original state.

On the one hand, ΔV generated in read operations should be high enough for the read sense amplifier and to reduce the soft error rate. As a result, a high C_{cell} is preferred. On the other hand, the high memory density requires that the physical dimension of the capacitor is as small as possible. In order to achieve high capacitance while retaining a small cell area, we can extend the capacitor design in DRAM to the third dimension. Figure 1.8 shows the cross section of a typical design of such a capacitor for DRAM. The trench capacitor is built through the diffusion area and deeply into the substrate. Its inner part is connected to the diffusion area of the transistor and the outer layer is connected to the substrate.

To build a DRAM array, the array size must be carefully chosen in order to balance the density and performance. A higher density can be easily achieved in a larger array because the peripheral circuitries, such as the sense

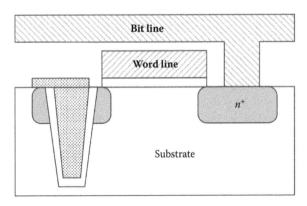

FIGURE 1.8
Cross section of the DRAM cell with a trench capacitor.

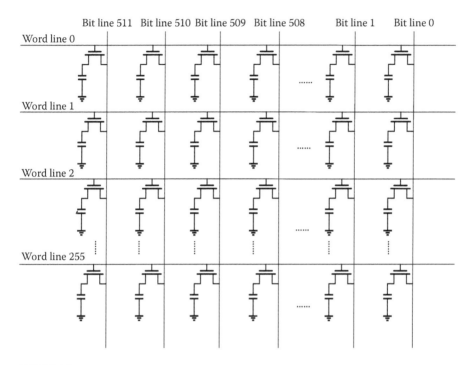

FIGURE 1.9
DRAM cell array.

amplifier and decoders, can be shared by more memory cells. However, the long bit line and word line in a large array also result in higher wire capacitances, which can degrade the access speed, lower the voltage swing, and increase the read error rate. A typical size of a DRAM subarray is 256 words by 512 bits, as shown in Figure 1.9.

The high storage capacity of DRAM makes it much cheaper than SRAM. Although its speed is slower than SRAM, it is preferred in many applications with high bandwidth requirements, such as the main memory in a computing system.

1.1.6 Nonvolatile Memories

Traditional memory technologies—for example, SRAM, DRAM, and flash memory—play a very important role in modern computing system and portable multimedia device industries. However, scaling of traditional memories poses significant technological challenges at 22 nm technology node and below due to process variations and device reliability [1, 2]. In recent years, significant efforts and resources have been put into research and development of emerging nonvolatile memory (NVM) technologies that combine features such as high density, high speed, low power, random accessibility, nonvolatility, and unlimited endurance. These characteristics allow NVM technologies to meet the requirements of various applications, from a large, expensive supercomputer to a low-cost, ubiquitous handheld device. Some promising candidates include phase change memory [3], magnetic memory (including both toggle-mode magnetic memory [4] and spin-transfer torque memory [5]), resistive memory, and memristors.

As we have discussed, ROM, EPROM, EEPROM and flash memory are all nonvolatile memories. However, in the rest of the book, we will mainly focus on emerging nonvolatile memory devices. We will introduce these technologies in detail in Chapters 2 to 6.

1.1.6.1 Phase Change Memory

Phase change memory (PCM) is fabricated with a chalcogenide alloy (typically, Ge_2-Sb_2-Te_5, GST) material, which is similar to those commonly used in optical storage means (compact discs and digital versatile discs) [3]. The data storage capability is achieved from the resistance differences between an amorphous (high-resistance) and a crystalline (low-resistance) phase of the chalcogenide-based material.

To switch a PCM cell to the crystalline phase, we can apply an electrical pulse to heat a significant portion of the cell above its crystallization temperature. When switching it to the amorphous phase, a greater electrical current is applied and then abruptly cut off in order to melt-quench the material [6].

The high resistance of a PCM cell can be four or five orders of magnitude higher than the low resistance, which makes MLC design possible.

1.1.6.2 Magnetic Random Access Memories

The basic storage component in MRAM is a magnetic tunneling junction (MTJ). Each MTJ has three layers: two ferromagnetic layers on the top and bottom and one tunnel barrier layer (made of MgO) in between as shown in Figure 1.10. The MTJ resistance is determined by the relative magnetization directions of the two ferromagnetic layers: when the magnetization directions of the two ferromagnetic layers are parallel (antiparallel), the MTJ is in a low (high)-resistance state. Data storage is realized by switching the MTJ between high- and low-resistance states [4].

In conventional MRAM design (known as *toggle mode*), MTJ resistance is changed by using the current-induced magnetic field to switch the magnetization of the MTJ. When the size of the MTJ scales down, the amplitude of the required magnetic field is increased correspondingly. The high write power consumption severely limits scaling of conventional MRAM. Recently, a new write mechanism based on spin polarization current-induced magnetization switching was introduced. This new MRAM design, called *spin-transfer torque RAM* (STT-RAM), is believed to have better scalability than conventional MRAM because its switching current is proportional to the MTJ size.

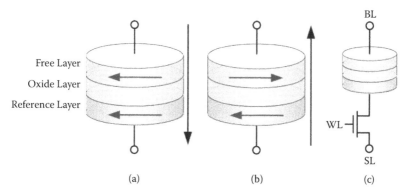

FIGURE 1.10
MTJ in (a) high resistance, (b) low resistance, and (c) a STT-RAM cell.

Various STT-RAM designs have been proposed by both industry and academia in recent years [5, 7].

1.1.6.3 Resistive RAM

Resistive RAM (R-RAM) generally denotes all memory technologies that rely on a resistance change to store information. In R-RAM, the data is stored as two (SLC) or more (MLC) resistance states of the resistive switching device. Resistive switching in transition metal oxides was discovered in thin NiO film decades ago [8]. Since then, a large variety of metal-oxide materials have been verified to have resistive switching characteristics, including TiO_2, NiO_x, Cr-doped $SrTiO_3$, PCMO ($Pr0.7Ca0.3MnO3$), CMO (conductive metal oxide), etc. The benefits of using a binary oxide system include a simple device structure and compatibility with conventional complementary metal–oxide semiconductor (CMOS) processing.

Based on the types of storage mechanisms, R-RAM materials can be cataloged as filament-based, interface-based, programmable metallization cell (PMC), etc. As we will introduce in Chapter 4, R-RAM has advantages such as nonvolatility, zero standby power, and high density.

1.1.6.4 Memristor

Nearly 40 years ago, Professor Chua predicted the existence of the memristor—the fourth fundamental circuit element—to complete the set of passive devices that previously included only the resistor, capacitor, and inductor [9]. The original definition of a memristor was derived from the completeness of circuit theory: in addition to the resistor, capacitor, and inductor, there must exist a fourth basic two-terminal element that uniquely defines the relationship between the magnetic flux (φ) and the electric charge (q) passing through the device [9], or:

$$d\varphi = M \cdot dq$$

Considering that magnetic flux and electric charge are the integrals of voltage (V) and current (I) over time, respectively, the definition of a memristor can be generalized as:

$$\begin{cases} V = M(\omega, I) \cdot I \\ d\omega / dt = f(\omega, I) \end{cases}$$

Here, ω is a state variable and M represents the instantaneous memristance, which varies over time [10].

In 2008, the first physical realization of a memristor was demonstrated by HP Lab in Palo Alto, California: the memristive effect was achieved by moving the doping front along a TiO_2 thin-film device [11]. Soon, memristive systems on spintronic devices were proposed [12].

The distinctive characteristic of a memristor to record the historic profile of the voltage/current through the memristor itself creates great potential in future system design. Its nonvolatility and excellent scalability make it a promising candidate as next-generation high-performance, high-density storage technology [13].

1.2 Devices

In this section, we will briefly introduce basic and common devices used in semiconductor memory chips, including MOSFET, diode, and passive components such as capacitor and resistor. Advances in semiconductor memory chip technology rely on the development of memory devices as well as these basic devices.

1.2.1 MOSFET

Semiconductor chips are built on a silicon substrate, and MOSFETs are widely used in order to achieve functionality.

According to the channel carrier type, there are two types of MOSFETs: (1) NMOS, whose D and S regions are formed by n^+ implantation and the carrier in the channel is electron; and (2) P-channel MOSFET (PMOS), whose D and S regions are formed by p^+ implantation, and holes serve as carriers in the channel. The circuit symbols of NMOS and PMOS are shown in Figures 1.11(a) and 1.11(b), respectively.

Usually NMOS transistors are built directly on p-type substrate, whereas PMOS transistors are fabricated inside N-wells in p-type substrates. The manufacturing process for building both NMOS and PMOS on the same substrate is called CMOS. CMOS technology has advantages such as high

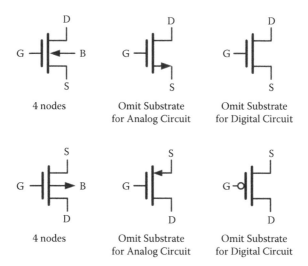

FIGURE 1.11
Symbols of (a) an NMOS transistor and (b) a PMOS transistor.

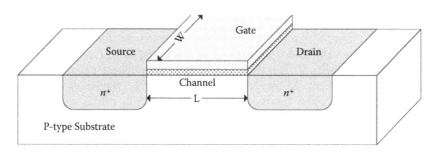

FIGURE 1.12
Cross section of an N-channel MOS transistor.

speed, low power consumption, and strong functionality and hence has been the dominant technology used in the semiconductor industry.

Figure 1.12 illustrates a cross section of an NMOS, which is fabricated on a p-type substrate. Through an ion-implantation process, two n^+ regions are formed in the substrate (B); one of them is called the source (S) and the other is called the drain (D). Note that because of the symmetric positions of the source and drain, these two regions are interchangeable, and we usually call the one with higher voltage the drain and the one with lower voltage the source. The channel can be formed right below the surface under the gate (G). The voltage of the gate controls the conductance of the channel. Usually the gate is made of polysilicon and is insulated from the channel by a thin silicon dioxide (SiO_2) between them, as shown in Figure 1.12.

The gate and drain voltages measured from the source are called V_{GS} and V_{DS}, respectively. As V_{GS} increases from 0 to a certain positive value (called

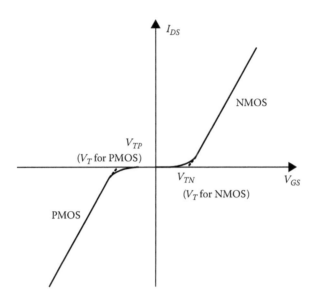

FIGURE 1.13
I-V curve of an MOS transistor.

the *threshold voltage* V_T), a channel of charge carriers can form under the silicon dioxide. The value of V_T is mainly determined by NMOS process technology. Once V_{GS} exceeds V_T and the $V_{DS} > 0$, there is a current (I_{DS}) flowing from the drain to the source. As illustrated in Figure 1.13, I_{DS} increases with V_{GS}.

Figure 1.14 shows the relation between I_{DS} and V_{DS} under different V_{GS}. We can divide these curves into three regions, as described in the following sections.

1.2.1.1 Cutoff Regions

In the cutoff region, V_{GS} is smaller than the threshold voltage V_T. As previously discussed, the channel between D and S is not formed in this situation, so $I_{DS} = 0$.

1.2.1.2 Triode Region

In the triode region, $V_{GS} > V_T$ and $V_{DS} < V_{GS} - V_T$. I_{DS} increases with V_{DS} by following the equation:

$$I_{DS} = \frac{W}{L}\mu C_{OX}\left[\left(V_{GS} - V_T\right)V_{DS} - \frac{1}{2}V_{DS}^2\right]$$

Here, W and L are the width and length of the channel, as shown in Figure 1.12; μ is the mobility of the carrier (electrons for NMOS and holes

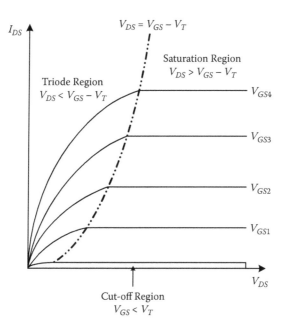

FIGURE 1.14
Relation between I_{DS} and V_{DS} under different V_{GS}.

for PMOS); and C_{OX} is the gate capacity per unit area. When $V_{DS} \ll V_{GS} - V_T$, $\frac{1}{2}V_{DS}^2$ can be omitted, and the equation becomes

$$I_{DS} = \frac{W}{L}\mu C_{OX}\left[\left(V_{GS} - V_T\right)V_{DS}\right]$$

At this time, I_{DS} is proportional to V_{DS}, so we can see the MOSFET at this region as a resistor, whose resistance is controlled by V_{GS}:

$$R_{DS} = \frac{1}{\dfrac{W}{L}\mu C_{OX}\left(V_{GS} - V_T\right)}$$

1.2.1.3 Saturation Region

When V_{DS} exceeds $(V_{GS} - V_T)$, I_{DS} stops following V_{DS} but maintains a relatively constant value:

$$I_{DS} = \frac{W}{2L}\mu C_{OX}\left(V_{GS} - V_T\right)^2$$

FIGURE 1.15
PN-junction structure.

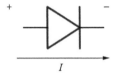

FIGURE 1.16
Symbol of a diode.

I_{DS} at this time is only determined by V_{GS}. More details and explanations for why an MOS transistor current becomes saturated when $V_{DS} > V_{GS} - V_T$ can be found in Razavi [14].

1.2.2 Diode

A diode is a two-terminal device that can conduct current only in one direction (called the *forward direction*) but blocks the current in the opposite direction (called the *reverse direction*). A semiconductor diode is made of a PN-junction by joining P-type and N-type semiconductors together in very close contact, as shown in Figure 1.15.

A symbol of the diode is shown in Figure 1.16. Here, the "+" and "–" represent the positive and negative terminals of the diode, respectively. The forward current is defined as the one flowing from "+" to "–", which can be calculated by the equation:

$$I = I_S \left[\exp\left(\frac{V_D}{nV_T} \right) - 1 \right]$$

where I is the diode current, I_S is the reverse bias saturation current, V_D is the voltage across the diode from "+" to "–", V_T is the thermal voltage, and n is the ideality factor, also known as the *quality factor* or the *emission coefficient*. In many cases, n is assumed to be approximately equal to 1.

The thermal voltage V_T is calculated by $V_T = kT/q$. Here k is the Boltzmann constant, T is the absolute temperature, and q is the magnitude of charge. At room temperature (300 K), V_T is approximately 26 mV.

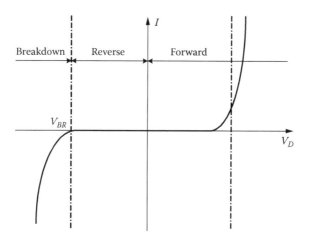

FIGURE 1.17
I-V curve of a diode.

Figure 1.17 illustrates the I-V curve of a diode. Note that when the reverse bias voltage reaches a certain point V_{BR}, the current through the diode increases significantly. This is called *breakdown* and will sometimes damage the device.

1.2.3 Passive Component

1.2.3.1 Capacitor

A capacitor is an important circuit component used in circuit design. In CMOS processes, there are several options to fabricate a capacitor. One method is to use polysilicon–insulator–polysilicon layers, which requires the addition of a second polysilicon layer and an extra oxide layer in between. The capacitance of $1fF/\mu m^2$ can be reached using this method.

Metal–insulator–metal layering is another more common capacitor realization. The corresponding cross section is shown in Figure 1.18. The insulator is usually between the top metal layer and the second metal layer in order to minimize the stray capacitance of the bottom plate. The unit capacitance of this type of capacitor is about $1~4fF/\mu m^2$.

We can also make a capacitor by using the gate and source/drain of a MOSFET, which can achieve a high capacitance per unit area. However, such a capacitor has nonlinear characteristics when operating in a large voltage range.

1.2.3.2 Resistor

Theoretically, a resistor can be fabricated at any layer in CMOS process technology. The resistivity of the material of the layer determines the resistance.

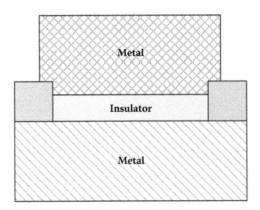

FIGURE 1.18
Cross-section of metal–insulator–metal capacitor.

Structures with relatively high resistivity (i.e., polysilicon and diffusion) are commonly used to build resistors. Because the diffusion area is usually associated with high parasitic capacitance to ground, it can be used only in low-frequency applications. When a polysilicon layer is used to build the gate in MOSFET, silicide is commonly used to reduce the gate resistance. In contrast, when it is used to make resistors, an extra mask might be needed to block the silicide on the area where the resistors are fabricated. Three hundred to 1,000 Ω/ð can be reached with a polysilicon layer.

In the resistor layer design, the meander structure shown in Figure 1.19 is commonly used to reduce the area cost.

1.3 Designing Memory and Array Structures

In this section, the general memory chip structure and some elementary circuit components will be introduced. Related topics such as the error correction and testability of a memory chip will also be included.

1.3.1 Memory Architecture and Building Blocks

Figure 1.20 illustrates a generic memory chip design, which can be divided into three major parts: a memory cell array, input/output (I/O) interface circuits, and peripheral circuits, including a decoder, sense amplifier, write driver, etc.

1.3.1.1 Memory Cell Array

The core of a memory chip is the memory cell arrays where the data is stored. A memory array is actually a matrix of the same memory cells.

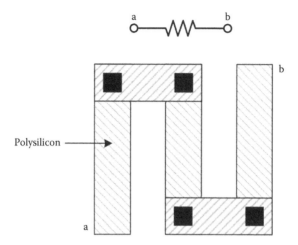

FIGURE 1.19
Meander structure during layout design of a resistor.

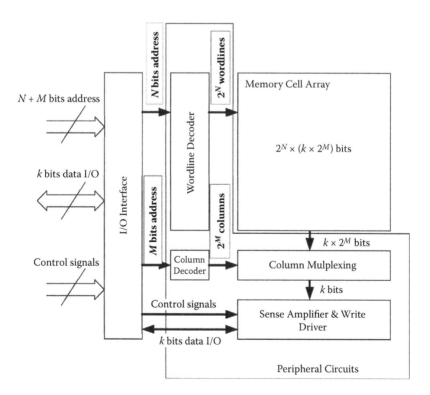

FIGURE 1.20
Memory chip structure.

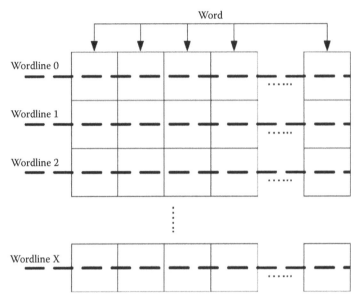

FIGURE 1.21
Cell array with each row containing only one word.

Usually, the cells in a memory array are accessed in groups, and each group is called a *word*. The cells in a word are located in the same horizontal row in the array. A word can contain 8, 16, 32, or even more bits, which depends on the system requirements. Due to the limitation of the number of pins in a chip package, the number of bits in a word cannot be too high: the more bits a word contains, the more pins are needed for the package of the memory chip. Compared to the number of bits in a word, the number of words in a memory array is much higher. Hence, if the memory array is designed as one word per row, it will become tall and thin, as shown in Figure 1.21. Such a design leads not only to high bit line load and poor performance but also highly complex word line decoder design.

In general, if a memory chip has 2^{N+M} words and each word has k bits, the memory array can be designed with a size of $2^N \times (2^M \times k)$ which means that it has 2^N rows and each row has $2^M \times k$ bits. An example of such a memory array is shown in Figure 1.22. To ensure that only one word is accessed at a time, a column decoder is needed to select k bits from $2^M \times k$ bits in a row at one time. The selection of M and N should consider the array capacity as well as the physical size of a single memory cell. The shape of the memory array is usually square to balance the capacitance loads on the bit line and the word line.

1.3.1.2 I/O Interface

The main function of I/O interface circuits is to convert the external input/output signals to the internal signals that can be accepted by the peripheral circuitry or vice versa.

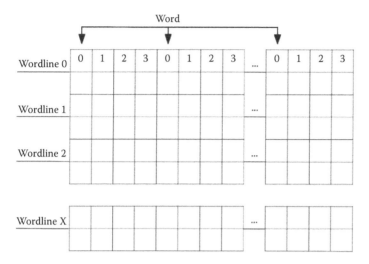

FIGURE 1.22

Cell array when each row contains four words (cells with the same number are from the same word).

For example, the I/O interface circuit receives N+M bits address and converts it into N+M pairs of complementary address signals (i.e., a_i and $\overline{a_i}$). N pairs are sent to a word line decoder and M pairs are sent to a column decoder. In write operations, the I/O interface circuit captures the input data and sends it to the write circuitry, whereas in read operations, it grabs the read-out data and sends them off-chip. The I/O interface also generates the internal control signals like precharge and enable signals for the sense amplifier according to the timing requirement from the external control signals.

1.3.2 Memory Peripheral Circuitry

The main peripheral circuits include word line/column decoder, sense amplifier, write driver, etc. Some peripheral circuits such as sense amplifiers can be shared by many memory elements to obtain high efficiency and high memory density.

1.3.2.1 Decoder

A word line/column decoder selects one word line/column with the given N/M bits address; that is, raising one word line/column-selecting signal to high voltage but keeping all the other selecting signals low. Thus, only the desired word line/column can be accessed during every memory access.

Figure 1.23 shows an example of a 2-to-4 decoder design, which sets only one of the four possible outputs to high with a given two-bit address (actually,

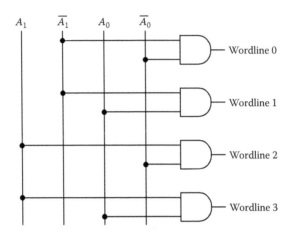

FIGURE 1.23
A 2 × 4 decoder.

two pairs of complementary addresses). Considering that each output of the decoder drives all of the selecting switches connecting to it, the load capacity is usually quite high. Therefore, in addition to carefully selecting a reasonable size of memory array, the drivability of the decoder needs to be carefully designed. Sometimes, a buffer might be needed between the decoder output and the selecting transistors. As the inputs of the decoder increase, the 1-level decoding scheme becomes too complex. The use of a pre-decoding circuit simplifies the design. For example, a straightforward 4-to-16 decoder needs at least 48 two-input gates (NAND and NOR). Figure 1.24 shows the alternative design with pre-decoding: First, it generates 8 internal pre-decoding signals, and these signals are used to produce the final word line signals. The new design needs only 24 two-input gates and hence significantly reduces the decoder design complexity.

The height of the word line decoder should perfectly match the corresponding memory array, which makes the layout design very tricky, especially when the dimension of memory cell is small; for example, DRAM.

1.3.2.2 Sense Amplifier

When reading out the data stored in a memory cell, different binary values can generate different voltages/currents at bit lines. However, the difference is usually quite small and cannot be used directly as output. A sense amplifier is used to amplify the difference so that it can be recognized properly by the I/O interface.

Figure 1.25 shows a latch-based sense amplifier design used in STT-RAM [7]. In STT-RAM design, the BL is connected directly to the memory cell and the source line (SL) can be set to a proper reference voltage value. According to the binary value stored in the cell, the BL can be either higher or lower

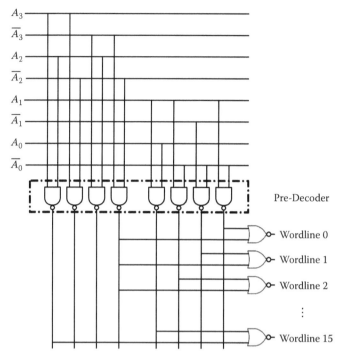

FIGURE 1.24
A 4 × 16 decoder with pre-decoding.

than the reference voltage. The positive/negative voltage difference between the BL and SL can affect the working point of the cross-coupled inverters at the top of Figure 1.25. The feedback loop of the cross-coupled inverters can further amplify the voltage difference and generate the corresponding digital outputs. In this design, both inputs and outputs of the sense amplifier need to be precharged before sensing in order to eliminate the influence from the previous sensing result. The sensing operation is activated by enabling the sense-enable signal (SAE).

The minimum input voltage difference that a sense amplifier can correctly function, called a *sensing margin*, is a crucial design factor. The sense margin must be smaller than the voltage swing generated by the memory cell to guarantee the correctness of read operations. Latency—that is, how long the input data can be sensed by a sense amplifier—is another important factor. It directly determines the read speed.

1.3.2.3 Write Driver

A write driver receives data from the I/O interface and then writes it into a memory cell. Figure 1.26 shows a write driver design used in STT-RAM [7], where the write current direction through the memory cell determines the value to be written. During a write operation, M1, M4, M5, and M8 are

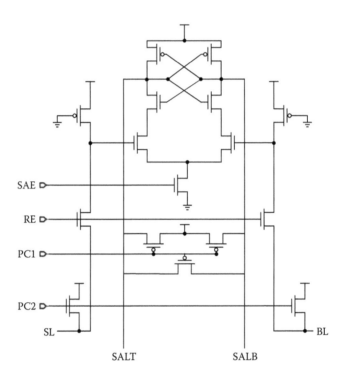

FIGURE 1.25
A latch-based sense amplifier [7].

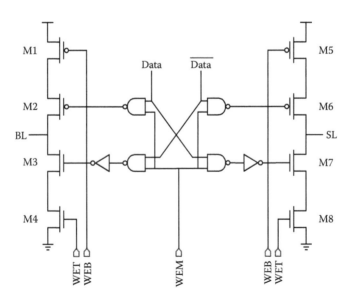

FIGURE 1.26
A write driver design for STT-RAM [7].

all turned on. When data '1' is to be written, M2 and M7 are turned on. Therefore, the PMOS transistors M1 and M2 behave as a current supply and the NMOS transistors M7 and M8 behave as a current sink. The write current flows from the current supply through a memory cell and ends at the current sink. The operation of writing '0' is similar to turning on M3 and M6.

1.3.3 Memory Reliability and Yield

1.3.3.1 Reliability

The reliability of a memory device is mainly evaluated by two parameters: write endurance and data retention.

Write endurance is characterized by how many times a memory cell can be programmed (written), which is determined by the memory cell structure, in particular the material for data storage. Different memory technologies have different endurances. For example, SRAM and DRAM can be programmed more than 3×10^{16} times. MRAM can achieve $\sim 1 \times 10^{16}$ write cycles [1]. The best demonstrated write cycle number is about 1×10^{12} times [1, 5]. The write endurance of flash memory is relatively low—1×10^5 times of erasing and writing up to operations [1]; this means that if a user rewrites the memory cell 10 times a day, the device can last nearly 27 years. Hence, flash memory is still good as a low-level storage device, such as a memory card for a digital camera. For some applications that require high endurance, wear leveling can be used to improve the reliability. Conceptually, wear leveling tends to evenly distribute the write operations into all of the memory cells. This prevents some memory cells from being heavily programmed and hence wearing out much faster than the other cells. Wear leveling techniques can effectively lengthn the lifetime of the overall memory chip [15].

Data retention defines how long data can be properly kept in a memory cell. Similar to write endurance, data retention is mainly determined by the memory process technology. For instance, the data in a DRAM cell is represented as the voltage level on the capacitor. As the charge leaks away, the data becomes invalid after a period of time; for example, ~ 1 ms. Thus, periodical refreshing is required in DRAM to maintain the data. Data retention can be an issue even in nonvolatile memory. For example, the retention time of MRAM is significantly impacted by the temperature of the environment: when the temperature increases from 300 to 350 K, the retention time of the MTJ (cell for MRAM) decreases from 6,000 years to 4 years [16]. In commercial products, the data retention time of nonvolatile memory is usually more than 10 years at room temperature.

1.3.3.2 Yield

The yield of memory chips is defined as the fraction of acceptable chips among all of the fabricated chips. This definition applies to all kinds of semiconductor chips. In memory design, yield can also specifically refer to the fraction of good memory cells among all of the memory cells in the design.

During the manufacturing process, many factors can cause defects in a memory chip, such as impurities in the wafer, dust particles on the masks, mask misalignment, etc. Moreover, fabricated devices, including memory devices and regular circuit components in CMOS technology, can deviate from their designed value. This can prevent the chip from working properly; for example, a transistor on a memory reading path may be designed to provide 50 μA read current but the device after fabrication supplies only 40 μA read current. The read current deviation may result in read-out data if it is not addressed at the design stage. Therefore, process variations should be properly considered during the design stage and some margin should be reserved in order to achieve a high memory yield.

Because of the high density and complicated production process, memory technologies are more vulnerable to physical defects than logic circuits. Some design techniques, such as redundancy and error correction codes (ECCs), which allow the chip to work even with some defective memory cells, are commonly used to improve the yield of memory chips.

Redundancy

The main purpose of the redundancy technique is to reserve some extra memory cells to replace defective ones when needed. Figure 1.27 illustrates an example of a redundancy technique with four spare rows [17]. The address comparators (AC0-AC3) can be used to replace defective word lines with spare word lines. During the testing stage, defective cells can be identified

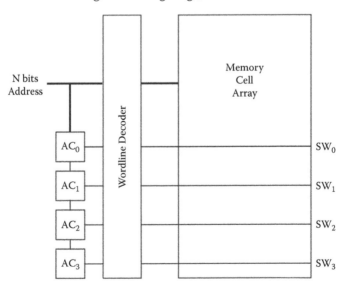

FIGURE 1.27
Redundancy technique with spare word line.

and then the address of these defective cells is programmed into the comparators. Whenever a comparator detects access to the defective word line, it switches to the corresponding spare word line and omits the original physical location.

Error Correction Codes

In general, error correction codes are error control systems for data storage, which systematically generate redundant data and store those redundant data together with the original data in memory devices. Carefully designed redundancy, which is added using a predetermined algorithm when the data are stored, allows the read circuit to correct a limited number of errors. The original data may or may not appear literally after the redundancy data are added using ECC algorithms (encoding) but, of course, they can be decoded back to the original data.

Commonly used ECCs include Hamming codes; Bose, Chaudhuri, and Hocquenghem (BCH) codes; etc. The interested reader can consult Micheloni et al. [18] for more details. Generally speaking, the stronger algorithms/codes provide a lower error rate. However, the complexity of the encoding/decoding circuit is also higher, which leads to more area overhead and higher latency. For a certain algorithm/code, adding more redundancy bits may aid in error detection and correction, but it also means greater area and higher chip costs. The designer should comprehensively consider the bit error rate, performance, cost, etc., to determine which ECC algorithm to use and how much redundancy to add.

1.3.4 Testing

Currently, the capacity of memory chips increases about four times every 3 years. Increasing memory density leads to longer testing time for each memory chip, which is directly related to the cost per chip. In order to keep memory prices economical, the testing time needs to be maintained within a reasonable scope. Thus, effective testing is increasingly important in producing successful memory chips. Many on-chip testing schemes have been proposed to reduce the testing time.

The purpose of some on-chip testing schemes is to quickly identify the defected chip, though they cannot provide the same high fault coverage as an external tester. Only the remaining chips that passed on-chip tests will be carefully tested by an external tester. Therefore, usage of the external tester can be reduced and hence testing costs can be decreased. Figure 1.28 shows one of these schemes, line-mode testing (LMT), which tests all of the memory cells on one selected line in one cycle [19]. Here, M bits of memory cells are read out and compared with the expected data stored in the registers. All of the comparison results are sent as inputs of an M-input AND gate. The test result can be '1' only when all of the comparison results are equal. The testing time of LMT is proportional to M for the conventional testing.

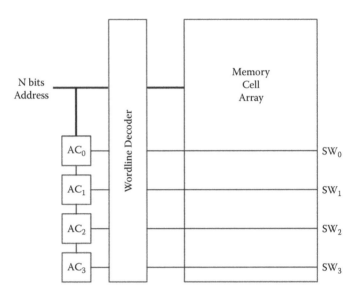

FIGURE 1.28
Line-mode testing scheme.

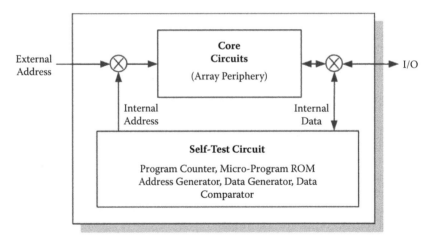

FIGURE 1.29
Built-in self-testing scheme.

A built-in self-test (BIST) has been widely used since the development of a system-on-chip (SOC), in which the memory is internally connected to the central processing unit (CPU) core and other components. In such a scenario, memory component testing with an external tester is difficult. As the name indicates, BIST can conduct the on-chip testing once the built-in self-test is activated, without the aid of external signals. An example BIST function is shown in Figure 1.29 [20]. This BIST model contains a program counter (PC), a micro-program ROM, an address generator, a data generator, and a data comparator.

References

[1] International technology roadmap for semiconductors. (2010). Available at: http://www.itrs.net/ (accessed February 28, 2010).

[2] Kim, K. and Jeong, G. (2007). Memory technologies for sub-40 nm node. Paper presented at the IEEE International Electron Devices Meeting (IEDM, Washington, DC).

[3] Bedeschi, F., Fackenthal, R., Resta, C., Donze, E., Jagasivamani, M., Buda, E., Pellizzer, F., Chow, D., Cabrini, A., Calvi, G., Faravelli, R., Fantini, A., Torelli, G., Mills, D., Gastaldi, R., and Casagrande, G. (2008). A bipolar-selected phase change memory featuring multi-level cell storage. *IEEE Journal of Solid-State Circuits*, 44(1): 217–227.

[4] Tehrani, S., Slaughter, J., Deherrera, M., Engel, B., Rizzo, N., Salter, J., Durlam, M., Dave, R., Janesky, J., Butcher, B., Smith, K., and Grynkewich, G. (2003). Magnetoresistive random access memory using magnetic tunnel junctions. *Proceedings of the IEEE*, 91(5): 703–712.

[5] Hosomi, M., Yamagishi, H., Yamamoto, T., Bessho, K., Higo, Y., Yamane, K., Yamada, H., Shoji, M., Hachino, H., Fukumoto, C., Nagao, H., and Kano, H. (2005). A novel nonvolatile memory with spin torque transfer magnetization switching: Spin-RAM. Paper presented at the IEEE International Electron Device Meeting (IEDM, Washington, DC).

[6] Burr, G., Kurdi, B., Scott, J., Lam, C., Gopalakrishnan, K., and Shenoy, R. (2008). Overview of candidate device technologies for storage-class memory. *IBM Journal of Research and Development*, 52(4–5): 449–464.

[7] Kawahara, T., Takemura, R., Miura, K., Hayakawa, J., Ikeda, S., Lee, Y. M., Sasaki, R., Goto, Y., Ito, K., Meguro, T., Matsukura, F., Takahashi, H., Matsuoka, H., and Ohno, H. (2008). 2 Mb SPRAM (spin-transfer torque RAM) with bit-by-bit bi-directional current write and parallelizing-direction current read. *IEEE Journal of Solid-State Circuits*, 43(1): 109–120.

[8] Gibbons, J. F. and Beadle, W. E. (1964). Switching properties of thin NiO flms. *Solid State Electronics*, 7(11): 785–790.

[9] Chua, L. (1971). Memristor—The missing circuit element. *IEEE Transactions on Circuit Theory*, 18(5): 507–519.

[10] Strukov, D., Borghetti, J., and Williams, S. (2009). Coupled ionic and electronic transport model of thin-film semiconductor memristive behavior. *Small*, 5(9): 1058–1063.

[11] Strukov, D. B., Snider, G. S., Stewart, D. R., and Williams, R. S. (2008). The missing memristor found. *Nature*, 453: 80–83.

[12] Wang, X., Chen, Y., Xi, H., Li, H., and Dimitrov, D. (2009). Spintronic memristor through spin-torque–induced magnetization motion. *IEEE Electron Device Letters*, 30(3): 294–297.

[13] Ho, Y., Huang, G. M., and Li, P. (2009). Nonvolatile memristor memory: Device characteristics and design implications. Paper presented at the IEEE/ACM International Conference on Computer-Aided Design, San José, CA.

[14] Razavi, B. (2001). *Design of analog CMOS integrated circuit.* New York: McGraw-Hill.

[15] Chang, L.-P. (2007). On efficient wear leveling for large-scale flash-memory storage systems. Paper presented at the ACM Symposium on Applied Computing (SAC), New York.

[16] Li, H., Wang, X., Ong, Z.-L., Wong, W.-F., Zhang, Y., Wang, P., and Chen, Y. (2011). Performance, power and reliability tradeoffs of STT-RAM cell subject to architecture-level requirement. Paper presented at the IEEE International Magnetics Conference.

[17] Eaton, S., Wooten, D., Slemmer, W., and Brady, J. (1981). A 100 ns 64 K dynamic RAM using redundancy techniques. Paper presented at the IEEE International Solid-State Circuit Conference (ISSCC).

[18] Micheloni, R., Marelli, A., and Ravasio, R. (2008). *Error correction codes for non-volatile memories*. New York: Springer.

[19] Nakagomem, Y. and Itoh, K. (1991). Reviews and prospects of DRAM technology. *IEICE – Transactions on Electronics*, 74(4): 799–811.

[20] Akira Tanabe, Toshio Takeshima, Hiroki Koike, Yoshiharu Aimoto, Masahide Takada, Toshiyuki Ishijima, Naoki Kasai, Member, IEEE, Hiromitsu Hada, Kentaro Shibahara, Takemitsu Kunio, Takaho Tanigawa, Takanori Saeki, Masato Sakao, Hidenobu Miyamoto, Hiroshi Nozue, Shuichi Ohya, Tatsunori Murotani, Kuniaki Koyama, and Takashi Okuda. (1992). A 30-ns 64-Mb DRAM with built-in self-test and self-repair function. *IEEE Journal of Solid-State Circuits*, 27(11): 1525–1533.

2

Phase Change Memory

Among all memory technologies introduced in this book, phase change memory (PCM) is probably the closest to the commercialization stage. Similarly, PCM has two or more distinct resistance states that represent the different logic values. Data storage, however, is based on switching between the amorphous phase with high electrical resistivity and the crystalline phase(s) with low electrical resistivity.

2.1 Introduction to PCM

2.1.1 Phase Change Properties

A phase-change material usually has two or more phases with quite different electrical properties, as shown in Figure 2.1. The materials can switch among these phases repeatedly and stably under some conditions; that is, under heating for certain lengths of time. As demonstrated by many phase-change materials, the amorphous and crystalline phases have significantly different optical and electrical properties; the amorphous phase features high resistivity and low reflectivity, whereas the crystalline phase shows low electrical resistivity and high reflectivity. Different optical and electrical properties can be used to represent the digital value; that is, 0 and 1.

2.1.2 History of PCM

Although PCM is categorized as a so-called emerging memory technology, the history of the phase changing concept may be longer than commonly thought. In 1962, Pearson et al. reported the switching phenomenon that exists in As-Te glasses [1]. In 1968, Ovshinsky described the fast transition (~10 ms) between the resistive and conductive states of some semiconductor materials by applying an electrical field [2]. This paper motivated considerable interest in the switching effects of chalcogenide materials; for example, a thin film of a noncrystalline alloy from Te, As, Ge, Si, etc. In 1969, Sie demonstrated a prototype of a PCM device by integrating chalcogenide film with a diode array [3, 4]. In 1970, Neale, Nelson, and Moore (later known as the cofounder of Intel) demonstrated the world's first 256-bit PCM [5]. Since then, the immature material

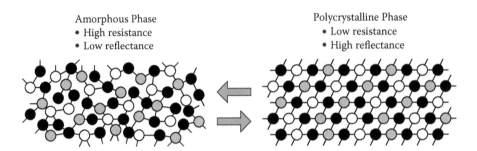

Amorphous Phase
- High resistance
- Low reflectance

Polycrystalline Phase
- Low resistance
- High reflectance

FIGURE 2.1
Phase changes between the amorphous phase and polycrystalline phase.

quality and high power consumption issues substantially slowed down further research and commercialization of PCM technology. However, PCMs (mainly chalcogenide) have been successfully used in the application of rewritable optical media, such as Compact Disc–Readable-Writable (CD-RW)/Digital Versatile Disc–Readable-Writable (DVD-RW), etc., during the last two decades.

Interest in PCM has been renewed recently for two main reasons: (1) the mainstream memory technologies, that is, flash, dynamic random-access memory (DRAM), and static random-access memory (SRAM) all face some scaling difficulties when the technology node enters the sub-100 nm range; and (2) some fast crystallizing materials such as GeTe-Sb_2Te_3 (GST) or Ag- and In-doped Sb_2Te (AIST), which can crystallize in about 100 ns, were discovered at the end of the 20th century [6, 7]. Therefore, the programming speed of PCM has the potential to be thousands of times faster than that of the existing nonvolatile memory like flash.

Competition for development of a commercial PCM product started around the transition to the new millennium. In 1999, Ovonyx was founded by Stanford R. Ovshinsky. The first modern PCM memory design was described in Wicker [8], partially based on his Ph.D. thesis [9]. In 2002, a 4-Mb test chip was fabricated with 0.18 μm 3 V complementary metal-oxide semiconductor (CMOS) technology as the first result of the collaboration between Ovonyx and Intel since the year 2000 [10]. Since then, the memory size of PCM has increased rapidly: In 2004, Samsung announced a 64-Mb PCM testing chip with 0.18 μm 3 V CMOS technology [11]; in 2006, Samsung set a record with 256 Mb [12]; in 2008, Samsung pushed its record to 512 Mb [13]; the latest record of a 1-Gb, 45-nm PCM chip was reported by Numonyx, a joint venture between Intel and STMicroelectronics [14].

2.1.3 Phase Change Operation

Figure 2.2 shows a conceptual cross section of a PCM cell. A layer of chalcogenide is integrated between a top electrode and a bottom electrode. A heating element (normally a resistor) is grown on the bottom electrode and contacts the chalcogenide layer.

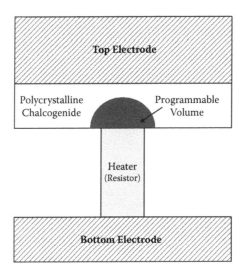

FIGURE 2.2
Cross section of a PCM cell.

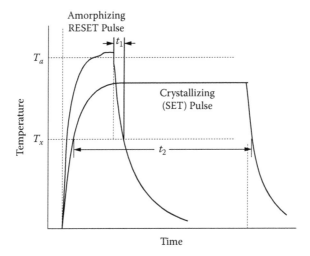

FIGURE 2.3
Programming procedures of a PCM cell [15].

The programming of a PCM cell can be divided into two different operations, as shown in Figure 2.3 [15]. During a set operation, a moderate but long current is injected into the heater and raises the temperature of the chalcogenide layer through Joule heating. Then the chalcogenide layer is crystallized by heating it above its crystallization temperature (T_x). The set pulse

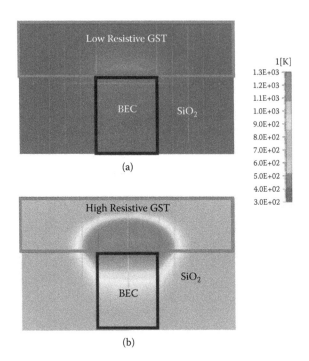

FIGURE 2.4
Simulation results of temperature distribution at the same reset (1 mA, 50 ns) pulses for two GST cells with different resistivities: (a) 2 mW-cm and (b) 20 mW-cm [19].

lasts for a sufficiently long duration ($t2$) to maintain the device temperature in the rapid crystallization range a time sufficient for crystal growth. The PCM cell shows a low-resistance state.

During a reset operation, a high but short current is injected and raises the temperature of the programmed volume in the chalcogenide layer above the melting point. The polycrystalline order of the material is eliminated and the volume returns back to the amorphous state. When the reset pulse is sharply terminated, the GST cell quenches to "freeze in" the disordered structural state. The device shows a high-resistance state. Because of the change in reflectivity, the amorphous volume appears as a mushroom cap-shaped structure. This quench time ($t1$) is usually determined by the thermal environment of the device and the falling time of the reset pulse (t_1).

The resistivity of GeSbTe (GST) film plays an important role in the operation of PCM. For example, Figures 2.4(a) and 2.4(b) [16] show the simulation results of temperature distribution at a PCM cell. When the same reset current pulse (1 mA, 50 ns) is applied, the temperature at the contact between the GST film and the heater can vary from ~141 to 973°C when the resistivity of GST films is 2 and 20 mW-cm, respectively. Properly increasing the GST film resistivity is important to reduce the operation power of PCM [17].

2.2 Material Research

2.2.1 $Ge_2Sb_2Te_5$ Alloys

Among all materials developed for PCM, chalcogenide alloys are the most commonly used and most widely studied. Chalcogenide alloys were originally invented for optical disk memory. Conventional chalcogenide alloys can be quickly (and easily) amorphized but require longer laser irradiation times (~1 ms) for crystallization. Therefore, conventional chalrogenide alloys were referred to as "a facility for amorphization" [6 (p. 2849)]. Such long programming pulse durations are certainly not in the interests of solid-state memory technologies. Yanada et al. [6] also found that the cooling rate of PCM is very fast when laser quenching is applied; for example, 10^{11} degrees/s. Such an extremely rapid cooling rate makes amorphization of any material possible. Therefore, instead of fast amorphization speed, fast crystallization speed (or a facility for crystallization) became a more favorable characteristic in the research on PCM.

Yamada et al. [6] also proposed the first such material, $GeTe-Sb_2Te_3$, a pseudobinary alloy belonging to the GST system. Figure 2.5 shows the composition transition among germanium (Ge), antimony (Sb), and tellurium (Te). When the composition of alloy changes from Sb_2Te_3 to GeTe, both the melting point and the data retention increase, and the crystallization speed decreases. Therefore, $GeTe-Sb_2Te_3$ provides a good tradeoff between the good cell stability of GeTe and the fast crystallization speed of Sb_2Te_3. Table 2.1 shows the thermal properties of different compositions of GST alloys, including crystallization temperature T_x, activation energy E_a, and amorphization temperature (melting point) T_m [18]. A higher E_a generally leads to higher T_x and T_a as well as better amorphous stability but leads to longer crystallization time, as shown in Table 2.1.

2.2.2 Doped GST

To achieve high crystallization speed and amorphous stability simultaneously, some dopants were added into the phase-change material. In [19], van

TABLE 2.1

Thermal Properties of Three Stoichiometric Compositions [18]

Compositions	T_x (°C)	E_a (eV)	T_m (°C)
$Ge_2Sb_2Te_5$	176	2.28	621
$GeSb_2Te_4$	158	2.12	617
$GeSb_4Te_7$	151	2.05	607
GeTe	196	2.69	725
Sb_2Te_3	—	—	620
$GeSbBiTe_4$	187	1.70	570

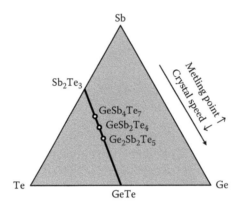

FIGURE 2.5
Alloy composition transition.

TABLE 2.2

Material Properties of Doped Sb-Te–Based Material Compositions [19]

Composition	Archival Life @ 50°C	Crystallization Time (Mark Radius of 125 nm) (ns)
Sb_2Te	—	—
$Ge_8Sb_{72}Te_{20}$	—	33
$Ag_8Sb_{72}Te_{20}$	100 Days	42
$In_8Sb_{72}Te_{20}$	1 h	29
$Ga_8Sb_{72}Te_{20}$	2 h	20
$Ga_8Sb_{77}Te_{15}$	0.5 h	—
$Sn_8Sb_{72}Te_{20}$	Less than a few days	17
$Ge_{15}Sb_{85}$	1×10^6 Years	15
$Ge_{22}Sb_{78}$	3×10^{14} Years	23

Pieterson et al. extensively analyzed the influence of various dopants on Sb-Te and Sb on some material parameters. For example, the archival life stability, which is determined by the activation energy for crystallization and corresponds to the data retention time of a PCM cell, decreases in the order Ge > Ag > Ga, In > Sn. Crystallization time, which is an essential parameter for PCM performance, decreases in the order Ag > Ge > In > Ga > Sn. The properties of various compositions are shown in Table 2.2.

Some doped GST films were also extensively studied. Cho et al. [19] found that the resistivity of GST film can be raised when nitrogen concentration increases, as shown in Figure 2.6. The improved resistivity of GST film helps to reduce the reset current. Also, N doping substantially increases crystallization temperature and improves the data retention of the corresponding PCM cell. One explanation is that nitrides are more covalent than GST. The

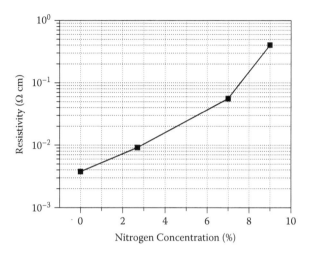

FIGURE 2.6
Resistivity of the GST films as a function of nitrogen concentration [16].

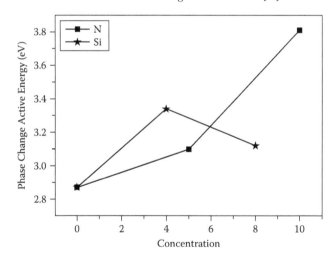

FIGURE 2.7
Phase change activation energy vs. dopant concentration [21].

presence of more covalent bonds reduces the atomic diffusivity and leads to an increase in the phase change activation energy [20].

Similar effects were found in Si-doped GST films, as shown in Figure 2.7 [21]. However, excessive Si doping will cause a decrease in the phase change activation energy of GST films, possibly because some crystal nuclei are formed in amorphous Si-doped GST materials and generate a large crystallization driving force.

Another doping option is bismuth (Bi), which can visibly reduce the reset current pulse width and amplitude with lower phase change activation energy

[18]. Bi is in the same periodic table element group as Sb (group 15) but has a larger radius. The bond length formed with other atoms is longer than Sb's, resulting in a weaker bond compared to Sb. During melting, bonds between Bi and other atoms are easily broken, resulting in a lower melting point.

2.2.3 GeSb Material

Although (doped) GST alloy has attracted the most attention in the development of PCM materials, some research has also been conducted on other materials. GeSb is one of the most well-known examples.

Sb content also affects some properties of Sb-based alloys: when increasing the Sb content, both the archival life stability and crystallization time decrease, as the last two rows in Table 2.2 show [19].

2.3 Device Research

The mushroom shape shown in Figure 2.2 is a conceptual structure of PCM cells. However, many advanced memory device structures have been studied in addition to the basic mushroom shape.

2.3.1 Bridge Cell

In Chen et al. [22], a PCM cell structure called phase-change bridge (PCB) was proposed based on N-doped GeSb materials. As shown in Figure 2.8, a PCB includes a narrow wire made of ultrathin phase-change material bridging two electrodes. Figure 2.8 also shows a potential integration scheme: one electrode is connected to the drain of the access transistor, and the other is connected to the source of the transistor and the bit line (BL). The electrodes are separated by a narrow oxide gap to obtain a reasonable threshold voltage.

Figure 2.9 shows the resistivity changes of doped GeSb ultrathin films when the temperature increases. They change from an amorphous state to a crystalline state at high temperature, nearly 100°C higher than that required by undoped $Ge_2Sb_2Te_5$ (GST). More important, the crystallization properties of GeSb thin film do not vary much even at an ultrathin film thickness of ~3 nm. This indicates good scalability of such a device structure and materials.

As shown in Figure 2.10, the working voltage of the PCB cell is below 1.4 V, which is within the normal working conditions of CMOS circuitry. The corresponding programming current is limited under 400 mA, which could possibly be supplied by a diode or a moderate-sized metal-oxide semiconductor (MOS) transistor.

Further investigation of Figure 2.11 shows promising cycle-by-cycle resistance variations and endurance performance: the on/off ratio, or the ratio between the high and the low resistance of PCB, keeps more than 10 after 30,000 programming (set/reset) cycles.

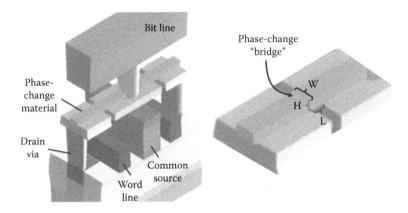

FIGURE 2.8
Phase change bridge cell structure and integration [22].

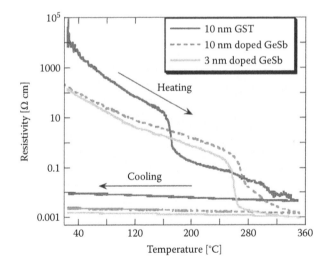

FIGURE 2.9
Crystallization behavior from amorphous-as-deposited ultrathin films of doped GeSb, compared to 10-nm-thick undoped Ge$_2$Sb$_2$Te$_5$ (GST) [22].

2.3.2 Pillar Cell

Although PCB cells demonstrate many promising memory characteristics, the bridge-based structure may require some manufacturing techniques that may not be fully compatible with current CMOS processes. Another cell structure, called a *pillar*, was proposed to maximize the usage of the current CMOS fabrication technology and minimize additional manufacturing costs [23].

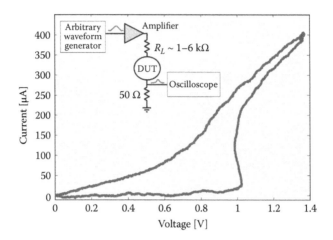

FIGURE 2.10
I-V characteristics of a PCM memory cell test structure [22].

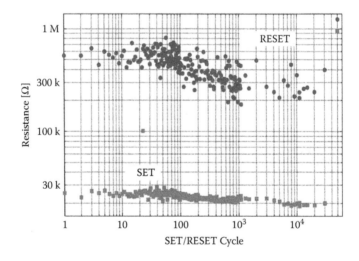

FIGURE 2.11
Cycle-by-cycle resistance variations and drifts after 30,000 programming cycles for PCB memory cell test structure (H = 3 nm, W = 20 nm, L = 50 nm) [22].

Figure 2.12 shows the pillar PCM cell structure, including a pillar of N-doped GeSbTe (GST:N) PCM sandwiched between the W plug connected to the access transistor and the Cu bit line. The electrically fully confined self-heating phase change regions result in lower reset currents compared to a conventional heater cell using an electrothermal finite element. As shown in Figure 2.13 [23], the reset current can be reduced by about half when the critical dimension (CD) of the process is scaled below 30 nm.

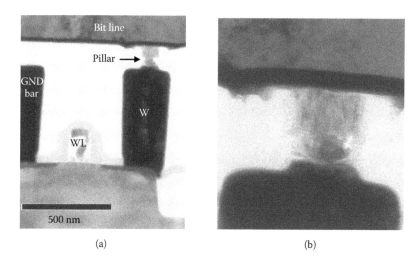

(a)   (b)

FIGURE 2.12
Pillar PCM cell structure: (a) TEM micrograph of a pillar memory unit cell with field-effect transistor (FET) access transistor and (b) TEM micrograph of a 75-nm pillar [23].

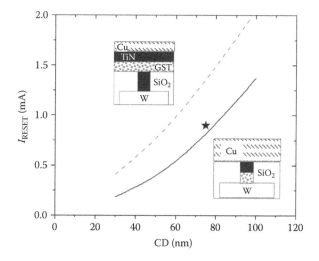

FIGURE 2.13
Simulated reset current for pillar and conventional heater cell structure at different CD sizes [23].

2.3.3 Pore Cell

Reducing the size of a PCM cell can effectively reduce the required set/reset current. Therefore, bridge and pillar PCM cell designs use so-called sublithographic technology to create feature sizes less than the critical dimensions. Therefore, a pore PCM cell was proposed to decouple the lithography

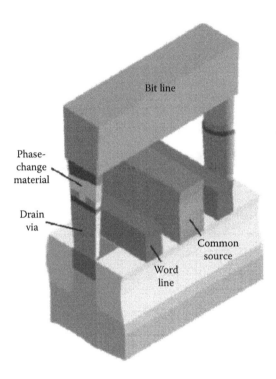

FIGURE 2.14
Schematic of two neighbored pore PCM cells [24].

variation associated with the sublithographic feature size. In the other words, the sublithographic CD becomes independent of the feature size defined by certain process technologies.

As shown in Figure 2.14 [24], a phase change element (PCE) is connected to an N-channel metal-oxide semiconductor field-effect transistor (NMOS) in series in a pore memory cell. A keyhole is formed with a sublithographic aperture.

Fabrication of the keyhole is illustrated in Figure 2.15, which can be summarized as follows: (1) a hole is first etched into an SiN-SiO$_2$-SiN stack using conventional lithography and stops on the bottom SiN layer; (2) a selective wet etch is performed to recess the SiO$_2$ layer, creating an overhang in the SiN layer; and (3) a highly conformal poly-Si film is deposited. This pinches off at the upper SiN layer and results in a keyhole in the poly-Si layer.

The testing results clearly demonstrate the independence of the electrical properties of the pore PCM cell on the original feature size of the process technology; say, the initial hole diameter. As shown in Figure 2.16, comparing the keyhole pore process with a (simulated) conventional spacer process, using the same initial lithographic hole sizes, the reset current dependence on the initial lithographic size is nearly eliminated.

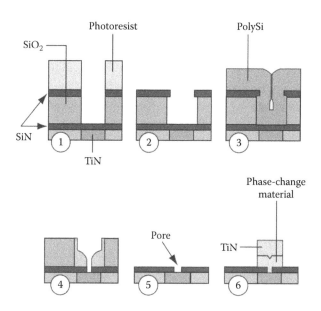

FIGURE 2.15
Process of sublithographic keyhole forming [25].

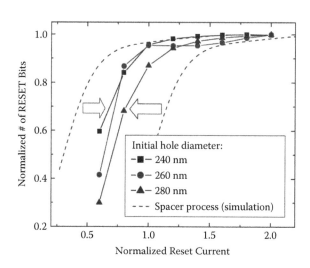

FIGURE 2.16
Reset current distributions for different initial hole diameters [24].

2.4 Electrothermal Modeling

It is well known that when current is applied to a resistor, part of the consumed energy will be translated to heat. In a PCM cell, a thermoelectric (or electrothermal) effect, which denotes the direct conversion of temperature differences to electric voltage and vice versa, is observed due to good electrical and poor thermal conductivity of a PCM: when a voltage is applied to a thermoelectric device, different temperatures will be generated on each side. In PCM design, this effect is used to heat the phase change device.

The major electrothermal effect in PCM operation is the Thomson effect, which was predicted and subsequently experimentally observed by William Thomson (Lord Kelvin) in 1851. It describes the heating or cooling of a current-carrying conductor with a temperature gradient.

If an electrical current density J passes through a conductor, heat production per unit volume is

$$q = \rho J^2 - \mu J \frac{dT}{dx} \tag{2.1}$$

where ρ is the material resistivity, dT/dx is the temperature gradient, and μ is the Thomson coefficient. The first term ρJ^2 is Joule heating and the second term is Thomson heating, which is reversible: it can be positive or negative when the current changes direction. The Thomson coefficient μ is equal to $T \cdot dS/dT$, where S is the Seebeck coefficient and T is the absolute temperature. In Castro et al. [26], the following differential equation was proposed to simulate the stationary temperature distributions in line-type PCM cells:

$$\lambda \frac{d^2 T'}{dx^2} - \mu J \frac{dT}{dx} - hT' + J^2 \rho = 0 \tag{2.2}$$

Here γ is the thermal conductivity. $T' = T - T_0$, where T_0 is the ambient temperature. hT' represents the heat loss to the surrounding materials. The coefficient h is determined by the geometry of the phase change line and surrounding materials, as well as their thermal conductivity. The boundary condition of Eq. (2.2) is $T'(0) = T'(L) = 0$, where L is the line length.

Thomson heating results in an asymmetrical temperature distribution along the PCM line, as shown in Figure 2.17. The location of the maximum temperature is deviated from the middle of the line, as the quantity δ shows.

Figure 2.18 shows the transmission electron microscopy (TEM) top view of a symmetrical PCM line under different electrode polarities. Shifts of the amorphous zone from the middle of the line are observed in both cases.

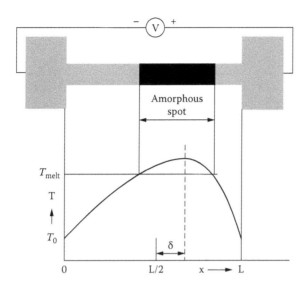

FIGURE 2.17
Asymmetrical temperature distribution along the PCM line due to the Thomson effect [26].

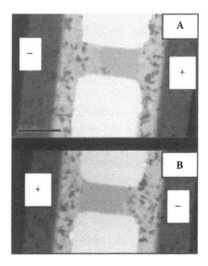

FIGURE 2.18
TEM top view of a symmetrical PCM line [26]. In (a) a shift of $\delta \sim 100$ nm of the amorphous zone toward the anode can be observed. In another cell (b), the amorphous mark is pushed against the other anode by using a 10% higher programing current.

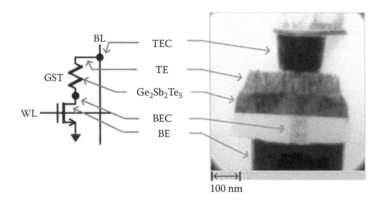

FIGURE 2.19
Basic PCM cell structure using transistors as the select devices [27].

2.4.1 PCM Design Technique

Following the maturity of PCM device development, many PCM designs have been proposed by leading semiconductor companies. Multilevel cell and other advanced design techniques have also been developed for the improvement of memory density and reliability. In this section, we will introduce some of these techniques.

2.4.1.1 Access Device of PCM Bit Cell

It is straightforward to connect the PCM device to an MOS transistor in series in a PCM cell, as shown in Figure 2.19 [27]. The existence of an NMOS transistor serves two purposes: first, it supplies the current to the PCM device; second, it can be used to select the PCM cells by turning it on or off. In the design shown in Figure 2.19, the top of the PCM device (GST) is connected to the BL through a top electrode contact (TEC). A bottom electrode contact (BEC) is formed below the GST and connected to the drain of access transistor terminal through a bottom electrode (BE). This simple cell structure can be easily integrated into the conventional CMOS process. Here the BEC works as a heater to realize the phase transition of GST between set and reset states. Here NMOS transistors are used due to their greater driving ability than P-type metal-oxide semiconductor (PMOS) transistors.

The corresponding 512-KB memory core using transistor as the select devices is shown in Figure 2.20. Because of their larger sizes, the word line drivers (SWD) are interleaved and put on two sides of the array to achieve higher integration density of the memory cells. Each two columns share one BL or one word line (WL), which is connected to the PCM devices. The details of the peripheral circuit will be discussed in the section on peripheral circuitry.

Although a transistor provides easy control of the memory cells, the current supplied by the transistor is greatly limited by the weak driving strength of

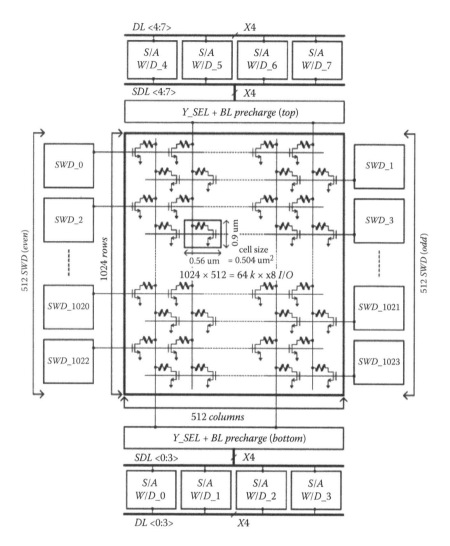

FIGURE 2.20
512-KB subarray core architecture [27].

MOS transistors. The cell size of the PCM cells is then primarily determined by the MOS transistor size instead of the PCM device. To reduce the memory cell size and improve the scalability of the PCM, diodes, which have much higher driving ability than MOS transistors, have also been proposed to supply the high programming current to PCM devices and melt the PCM.

Figure 2.21 shows a diode-switch PCM cell design proposed in K.-J. Lee et al. [28]. A vertical diode is grown below the PCM cell by using selective epitaxial growth (SEG) technology. The vertical PCM cell structure includes a BL, a TEC, a GST material, a self-aligned BEC, a vertical SEG P-N diode,

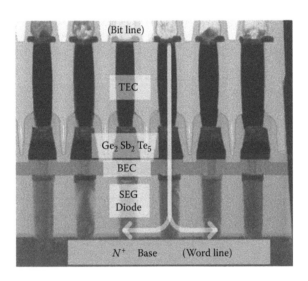

FIGURE 2.21
Vertical view of a diode-switch PRAM cell [28].

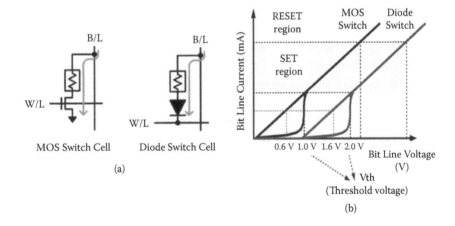

FIGURE 2.22
PRAM cells: (a) schematic views of PCM cells using an NMOS transistor or diode as the select device and (b) I–V characteristic curve comparisons [28].

and a WL built on an N+-doped base. The top surface area of a PCM cell can reach the theoretical minimum size of $1F^2$, which is constrained only by the feature size of a certain process technology node.

A diode-switch PCM cell has a higher operating voltage than an MOS-switch PCM cell (about 1 V higher) due to the higher threshold voltage, as shown in Figure 2.22. When using an NMOS transistor as the selected device, the reset current is always applied from the BL to the ground because of the high V_{GS} of the transistor. In diode-based design, the voltage applied

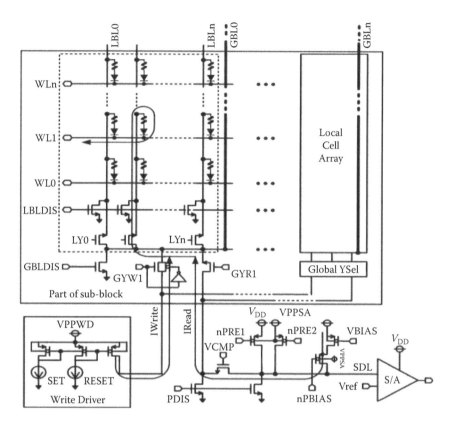

FIGURE 2.23
Memory core design for diode-switch PCM [28].

must be high enough to make the diode conductive. Therefore, a higher BL voltage must be applied in diode-based design to achieve the same programming current supplied by an MOS transistor. However, the surface area of the diode is still much smaller than the transistor area in such a case.

Figure 2.23 shows the corresponding memory core design for the diode-switch PCM. Deselected WLs are set to V_{DD} + 1–2 V to turn off deselected diode switches. Deselected BLs are floating to reduce current leakage and parasitic effects during normal write and read operations. To turn on a selected diode switch, the selected SL is pulled down to ground (GND). Write current I_{write} and read current I_{read} are supplied by write drivers or a PMOS transistor according to the operating mode.

In addition to conventional diodes, some advanced select device fabrication technologies have been investigated; for example, a Ge nanowire diode [29]. Using a nanowire as the select device may reduce the bottom electrode contact size due to the small diameter of the nanowire.

Figure 2.24 shows such a PCM cell structure proposed in Zhang et al. [29]. A phosphorus-doped germanium nanowire (GeNW) was used as the BEC.

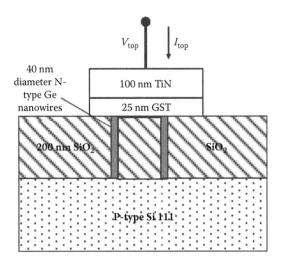

FIGURE 2.24
Schematic structure of a PCM cell using a Ge nanowire diode as the select device [29].

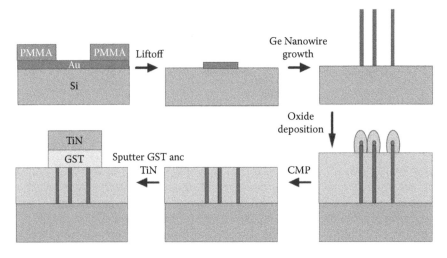

FIGURE 2.25
PCM cell fabrication flow [29].

The n-doped nanowire formed a p-n junction with the p-type substrate. The size of the nanowire was determined by the catalyst volume and growth condition and therefore could be below the lithographic limit. Similar to conventional diodes used in K.-J. Lee [28], the threshold voltage was between 1 and 2 V.

The fabrication process is shown in Figure 2.25. Poly(methyl methacrylate) (PMMA) is patterned by e-beam lithography on the top of p-doped (boron) Si(111) substrate. A thin Au layer is then evaporated followed by PMMA

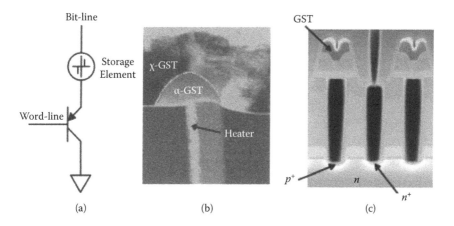

FIGURE 2.26
(a) Schematic of a PCM cell using a bipolar transistor as the select device; (b) scanning electron micrography (SEM) picture (along the BL direction) of a PCM cell, including the crystalline (x) and the amorphous (α) GST and the heater; (c) detail of the array along the WL direction (P⁺: BJT emitter; n: BJT base region; n⁺: base contact). The collector region of the BJT select device, which corresponds to the common substrate, is not included in the figure [30].

liftoff as a GeNW catalyst. Then the GeNWs are grown on Si substrate and their orientation and density are controlled. In situ doping is obtained under a constant flow of PH_3 precursor during growth. After GeNWs are grown, dielectric (SiO_2) is deposited between GeNWs by plasma-enhanced chemical vapor deposition (PECVD). Chemical–mechanical polishing (CMP) is applied to polish the wafer surface so that a good contact can be formed between the top surface of nanowires and the GST device.

Some research efforts are also made using a bipolar junction transistor (BJT) as select device, though this is not fully compatible with conventional CMOS processes. Figure 2.26 shows the corresponding cell structure: The cell selection is realized by controlling the SL, which is connected to the base of the *pnp*-BJT device. The BJT emitter is connected to the bottom electrode of the PCM device—that is, the bottom terminal of the heater—through a tungsten precontact. The BJT collector is formed by the common ground (chip substrate). The top electrode of the PCM device—that is, the top surface of the GST layer—is connected to the BL, which runs orthogonally to the WLs. As shown in Figure 2.26(b) and (c), the chalcogenide material is capped with a TiN barrier and deposited inside a sublithographic trench. The trench cell fabrication adopted a self-aligned (SA) technique to reduce the lithographic constraints.

Figure 2.27 shows that the BJT selector is able to deliver the 300 mA reset current at 1.5 V. Bedeschi et al. [30] reported that this was sufficient to supply the required programming current of the developed PCM device. It is worth mentioning that a very low base-emitter (BE) current leakage occurs under reverse bias conditions (less than 10 pA at a temperature of 85°C). The low

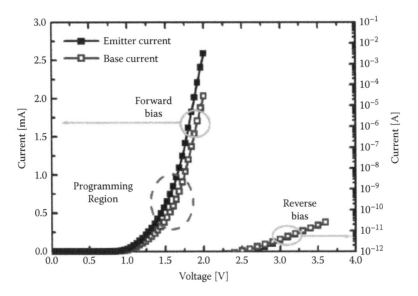

FIGURE 2.27
I-V characteristic of the *npn* bipolar selector under forward (left) and reverse (right) bias conditions [30].

leakage is necessary to build a very high-density array because it reduces disturbance in read operations.

Figure 2.28 shows the scheme to apply the voltage on the memory array during read and write operations. In normal operation of bipolar-switched PCM designs, the selected and unselected WLs are grounded or connected to a high voltage, respectively, to keep the corresponding selectors in the off state. The applied voltage of the selected BL is different in read and write operations (1.3 and 3.8 V, respectively) and, hence, the corresponding voltage applied to unselected WLs is also different (1.3 and 3.8 V, respectively). During reading, or programming, a multiplexing circuit connects each unselected BL to a virtual-ground line ($V_{HLV} = 0.3$ V).

2.4.1.2 Multilevel Cell

A multilevel cell (MLC) is an effective technology to boost memory density without increasing the memory cell area. In PCM design, 2_N resistance states can be used to denote four combination states of N digital bits; for example, 00, 01, 10, 11 for a two-bit MLC PCM cell. Because the resistance of the PCM material is determined by the amplitude and the duration of the applied current/voltage, it is possible to program the PCM cell to multiple resistance states by controlling the programming signals.

In Happ et al. [23], multiple PCM cell resistance states were programmed using two different approaches. In the first approach, the cell was programmed from the reset state using a single 300-ns set pulse with different

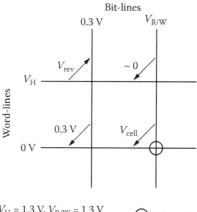

FIGURE 2.28
Voltage applying scheme of the memory array during read and write operations (arrows specify current direction). V_{cell} is the voltage across the storage element of the selected cell; V_{rev} indicates the presence of a reverse bias voltage [30].

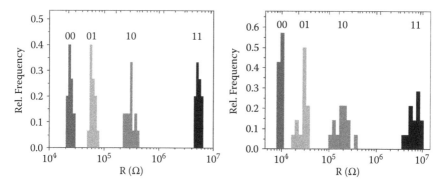

FIGURE 2.29
Histogram of MLC PCM cell resistance states: (a) using single set pulse from a reset state and (b) using a single reset pulse from a set state [23].

voltage magnitudes for different resistant state programming. In the second approach, the cell was programmed from the set state using a single reset pulse with different voltage magnitudes for different resistant state programming. Both approaches can achieve MLC functionality with four clearly differentiated resistance states, as shown in Figure 2.29. The intermediate states show slightly more widespread distributions compared to the two boundary states.

However, MLC PCM cells demonstrate significant temperature dependency, as shown in Figure 2.30. When the temperature increases, the PCM cell resistance drifts down and results in smaller differentiations between

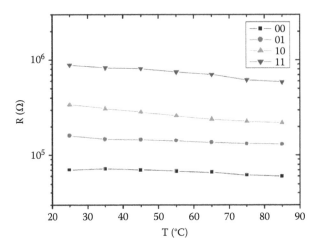

FIGURE 2.30
Temperature dependency of MLC PCM cell resistance [23].

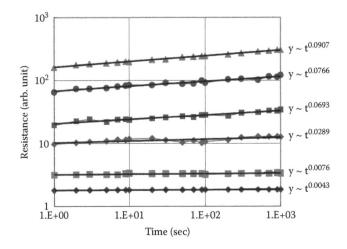

FIGURE 2.31
Temporal increases of PCM resistance levels at room temperature [31].

different resistance states. This resistance shift may cause potential errors during the read operations.

As we discussed before, a diode or bipolar transistor can supply greater driving strength with a smaller device area. MLC technology can also be applied to PCM designs using a diode or bipolar transistor as the select device.

Kang et al. [31] systematically investigated the operation of diode-switch MLC PCM designs. The test results in Figure 2.31 show that the resistance levels of MLC PCM cells increase at room temperature, where it is known that the PCM device resistance shift follows the empirical equation $R \sim R_i t^d$.

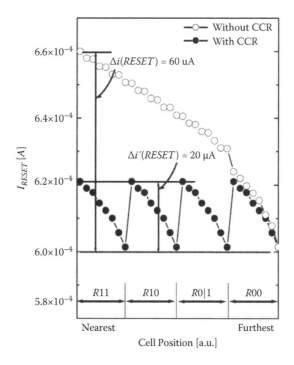

FIGURE 2.32
Reset current variations before/after using CCR [33].

Here R_i is the initial resistance right after a write operation. This resistance shift will help to suppress the error rate at the beginning due to the greater differentiation between the resistance levels and then cause a severe sensing error because the resistance significantly deviates from the normal sensing region [32].

Finally, MLC PCM cells can be driven by bipolar transistors, as demonstrated in Bedeschi et al. [30]. To ensure the sharp distribution of every resistance state, some novel programming techniques and algorithms were proposed. Details of the programming circuit for MLC PCM cells will be discussed in the next section.

2.4.2 Program Algorithm

The introduction of MCL technology in PCM design significantly increases the control complexity of PCM cell programming. As we mentioned in the last section, a simple MLC programming technique that changed only the applied voltage was proposed in Happ et al. [23]. However, to ensure successful sensing in multiple resistance states, the distribution of each resistance state must be wide enough to avoid large overlaps. Many novel write mechanisms were introduced to achieve this goal.

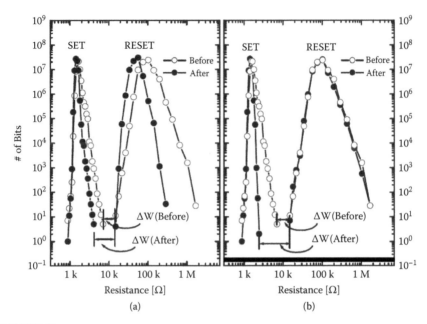

FIGURE 2.33
PCM cell resistance distribution before/after using (a) CCR and (b) MSPG [33].

The first problem in PCM cell programming is the unbalanced programming current to the memory cells at different locations. The voltage drop due to the interconnect resistance cause the memory cells that are far from the write drivers to suffer from a low write current. This becomes severe during reset operations due to higher programming current. When the reset current is targeted for far-end cells, however, the near-end cells will be over-programmed due to the excessive reset current.

In On et al. [33], a cell current regulation (CCR) scheme was proposed to adaptively adjust the reset current to the PCM cells. The memory cells were divided into multiple groups depending on their locations. Different levels of reset currents were then applied accordingly to minimize the current distributions: After applying the CCR scheme, the maximum cell current difference was reduced from 60 to 20 mA, as shown in Figure 2.32.

Another problem is the cell-to-cell process variations that exist in the PCM cell design; for example, the variations of optimal set current windows due to fluctuations of the contact area between the heater and PCM device and the composition ratio of the GST compound. A so-called multiple stepdown pulse generator (MSPG) technique is used to cover the cell-to-cell variations of set current windows by sweeping the magnitude of the set current [33] to achieve better PCM cell resistance distributions.

The applications of CCR and MSPG greatly improve the PCM device resistance variation. As shown in Figure 2.33, the differentiation between the set and reset resistance distribution (DW) is improved significantly.

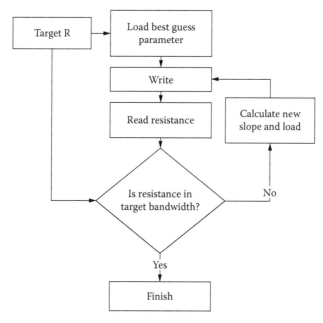

FIGURE 2.34
Programming pulse slope tuning flow for MLC PCM cells [34].

Other write strategies have also been proposed for MLC PCM cells based on the dependence of crystallization on temperature and time duration for which the programming current/voltage is applied. In Nirschl et al. [34], the falling slope of the programming current was adaptively set as a linear line or staircase by monitoring the PCM device resistance, as shown in Figure 2.34. Hence, the PCM device resistance could be precisely controlled.

By using this technique, up to four-bit MLC PCM cell design can be realized with sharp PCM cell resistance distributions. Figure 2.35 shows an example of the narrow resistance distributions of 100 (10 × 10 array) four-bit (16-level) MLC PCM cells using the proposed programming flow in Figure 2.34.

A program-and-verify (P&V) technique can also be applied to ensure adequate control of the PCM cell resistance, as shown in Figure 2.36 [30]. Each program current pulse is well controlled: The cell is first programmed to its set state by a proper set pulse, followed by a long set sweep initializing to its lowest resistance state. Then a single reset pulse with a fast quench is applied to initialize the cell to a reset state. All program pulses $I_{PGM,i}$ have the same pulse width, and the amplitudes increase step by step until the desired resistance is achieved.

2.4.3 Read and Other Peripheral Circuitry

In addition to write circuits, read circuitries are essential in PCM design to fulfill memory functions.

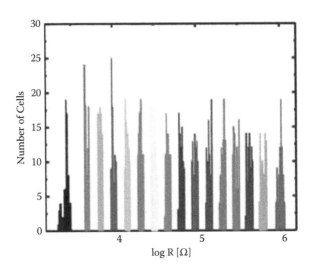

FIGURE 2.35
Resistance distribution of four-bit MLC PCM cells [34].

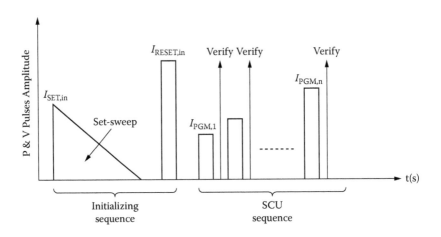

FIGURE 2.36
Program-and-verify MLC programming technique [30].

Due to the large differentiation between the different resistance states of PCM cells, some prototyping chips use a very simple sense amplifier (S/A) design [27]: After applying a read current to the PCM cells, the generated BL voltage will be directly compared to a reference voltage, as shown in Figure 2.37. However, the performance of such a design may be very slow if the voltage generated is significantly lower than V_{DD}.

A charge-transfer sense amplifier was used in PCM to sense the small signal and speed up the read performance [35]. As shown in Figure 2.38, three

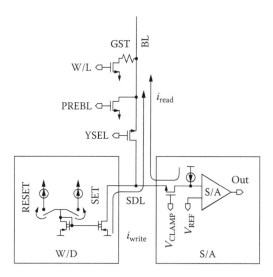

FIGURE 2.37
Simple sense amplifier design in PCM [27].

pairs of MOS transistors, (PA, PA'), (PB, PB'), and (NA, NA'), were included in the charge-transfer sense amplifier design. The first pair precharged the output nodes Sense Amplifier Input (SAINT), Sense Amplifier Input Bar (SAINB) to the V_{DD} supply while the BLs were precharged to a read BL voltage V_{BR}. When the second transistor pair was activated by the third pair, the output nodes were discharged. Because the readout voltage of BLs is sent to both the gate and drain nodes of the second pair of transistors (PB, PB'), the driving strength difference for this pair of transistors is increased. The overall sensing process can be completed within 10 ns.

In a design using bipolar transistors as the select device, the sense amplifier must be designed to compensate for the effect of the bipolar transistor along the sensing path, as shown in Figure 2.39 [30]. For example, the left branch mimics the BJT selector and the BL path of the array cell. The PMOS transistor in this branch is a dummy element that compensates for the on-resistance of the BL selector.

High programming voltage is used in diode- and BJT-based designs due to their relatively higher threshold voltage. Therefore, a charge-pump system might be needed in such designs, as shown in Figure 2.40. The basic voltage level V_{PPSA} for write and read operations is generated in both working mode and standby mode. However, in standby mode, the oscillation period of the charge pump is prolonged to meet the target standby power consumption.

2.4.4 Physical Limit of PCM

In addition to the basic memory operations such as read and write, there are many other fundamental properties that affect the usage of PCM; for

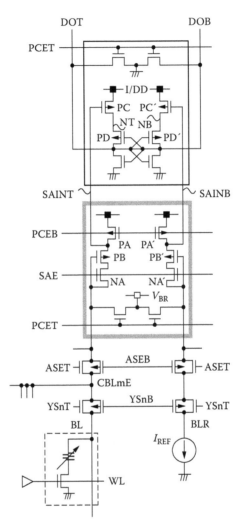

FIGURE 2.38
Charge-transfer direct-sense amplifier [35].

example, bit retention time and bit endurance. The retention time is defined as how long the memory cell can keep the data stable when the power is off; the endurance time is defined as how many cycles the memory cell can be programmed for. In this section, we discuss the retention time and endurance performance of PCM and the scaling property in future technology nodes.

2.4.4.1 Bit Retention

The structural relaxation process is the main factor affecting the bit retention time of a PCM cell, which is caused by the nature that chalcogenide material tends to minimize its free energy.

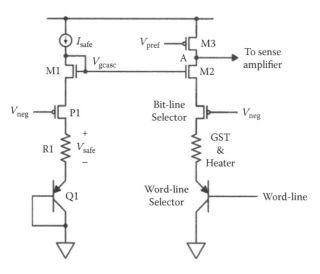

FIGURE 2.39
Read circuit design using bipolar transistors [30].

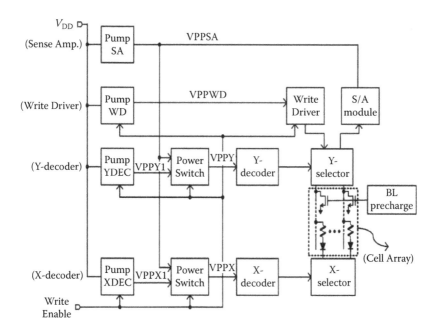

FIGURE 2.40
Charge-pump design in diode-based PCM designs [28].

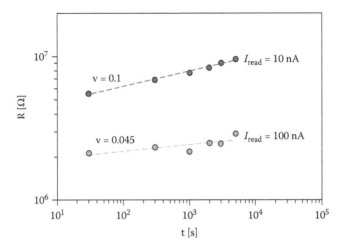

FIGURE 2.41
PCM device resistance as a function of time [36].

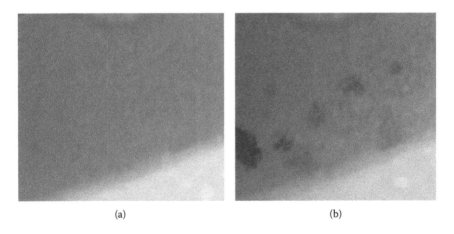

FIGURE 2.42
TEM images (a) before and (b) after e-beam (120 KeV) heating show spontaneous nucleation and grain growth in as-deposited GST film [40].

Structural relaxation results in a steady increase of resistance over time: the localized state density that is important to Poole–Frenkel transport in the amorphous semiconductor decreases followed by structural relaxation. In Lacaita and Jelmini [36], the measurement results showed that the PCM device resistance drift is reduced when the read current is increased, as shown in Figure 2.41.

Although the retention loss of a normal PCM cell is due to the grain growth from the amorphous/crystalline GST (aGST/cGST) boundary,

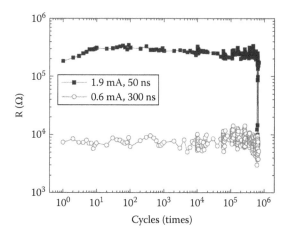

FIGURE 2.43
Endurance testing of a PCM cell [37].

further investigation has shown that the dominant mechanisms of the retention time degradation of the PCM cells at the tail of the distribution, which are defined as the PCM cells with very short data retention time under a high-temperature bake, are spontaneous nuclei generation and grain growth inside the GST active region [36], as shown in Figure 2.42.

2.4.4.2 Bit Endurance

After many write/erase (W/E) cycles, a PCM cell ends up with either reset-stuck failure (RSF) or set-stuck failure (SSF) [37]. In either mode, a PCM cell can no longer be programmed to the other state. In general, SSF is more common than RSF.

The SSF of a GST device was reported to be the result of Ge depletion and Sb enrichment inside the active volume of GST [38]. Detailed analysis indicated that SSF is mainly caused by field-induced migration of the PCM components. Thus, the lifetime of a PCM device can be extended by utilizing reverse migration induced by a reverse electric field. Figure 2.43 shows that a PCM device has a stable resistance value until it reaches the SSF after 6.9×10^5 cycles.

2.4.4.3 Continue Moore's Law

When the NAND flash technology enters the 28-nm range and beyond, providing comparable or better memory density using PCM becomes a very challenging task. A diode-based select device (plus cross-bar structure) is therefore a must to achieve $4F^2$ memory cell area, which has been achieved in NAND flash design. Moreover, the diode must be able to provide sufficient programming current to the PCM device.

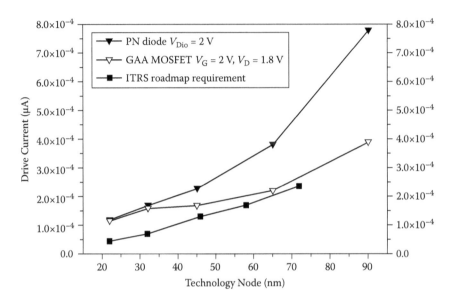

FIGURE 2.44
Driving ability of a PN diode and GAA MOSFET with technology scales [39].

Figure 2.44 shows the driving ability of PN diode and Gale-All-Around (GAA) metal-oxide semiconductor field-effect transistor (MOSFET), a new silicon-on-insulator (SOI) device with a symmetrical double-gate structure. For comparison purposes, the programming current of PCM devices predicted by the International Technology Roadmap for Semiconductors (ITRS) is also shown in the figure. Sufficient current can always be provided by the two technologies.

However, other performance features, such as bit retention time and endurance, are degraded when the technology scales [37]. Many technology obstacles must be overcome to replicate the success story of NAND flash memory in PCM technology.

References

[1] Pearson, A. D., Northover, W. R., Dewald, J. F., and Peck, W. F., Jr. (1962). *Advances in glass technology*. Plenum Press: New York.

[2] Ovshinsky, S. R. (1968). Reversible electrical switching phenomena in disordered structures. *Physical Review Letters*, 21(20): 1450–1453.

[3] Sie, C. H. (1969). Memory devices using bistable resistivity in amorphous As-Te-Ge films. Ph.D. dissertation, Iowa State University, Ames, IA.

[4] Pohm, A. V., Sie, C. H., Uttecht, R. R., Kao, V., and Agrawal, O. (1970). Chalcogenide glass bistable resistivity (ovonic) memories. *IEEE Transactions on Magnetics*, 6: 3.

[5] Neale, R. G., Nelson, D. L., and Moore, G. E. (1970). Nonvolatile and reprogrammable, the read-mostly memory is here. *Electronics*, 43(20): 56–60.

[6] Yamada, N., Ohno, E., Nishiuchi, K., and Akahira, N. (1991). Rapid-phase transitions of GeTe-Sb$_2$Te$_3$ pseudobinary amorphous thin films for an optical disk memory. *Journal of Applied Physics*, 69(5): 2849–2856.

[7] Tominaga, J., Kikukawa, T., Takahashi, M., and Phillips, R. T. (1997). Structure of the optical phase change memory alloy, Ag–V–In–Sb–Te, determined by optical spectroscopy and electron diffraction. *Journal of Applied Physics*, 82(7): 3214–3218.

[8] Wicker, G. (1999). Nonvolatile, high density, high performance phase change memory. *Proceedings of SPIE*, 3891: 2–9.

[9] Wicker, G. (1996). A comprehensive model of submicron chalcogenide switching devices. Ph.D. Dissertation, Wayne State University, Detroit, MI.

[10] Gill, M., Lowrey, T., and Park, J. (2002). Ovonic unified memory—A high-performance nonvolatile memory technology for stand-alone memory and embedded applications. Paper presented at the International Solid-State Circuits Conference, San Francisco, CA.

[11] Cho, W. Y., Cho, B., Choi, B., Oh, H., Kang, S., Kim, K., Kim, K., Kim, D., Kwak, C., Byun, H., Hwang, Y., Ahn, S., Jung, G., Jung, H., and Kim, K. (2004). A 0.18 µm 3.0 V 64 Mb non-volatile phase-transition random-access memory (PRAM). Paper presented at the International Solid-State Circuits Conference, February.

[12] Kang, S., Cho, W., Cho, B.-H., Lee, K.-J., Lee, C.-S., Oh, H.-R., Choi, B.-G., Wang, Q., Kim, H.-J., Park, M.-H., Ro, Y.-H., Kim, S., Kim, D.-E., Cho, K.-S., Ha, C.-D., Kim, Y., Kim, K.-S., Hwang, C.-R., Kwak, C.-K., Byun, H.-G., and Shin, Y. S. (2006). A 0.1 µm 1.8 V 256 Mb 66 MHz synchronous burst PRAM. Paper presented at the International Solid-State Circuits Conference, San Francisco, CA.

[13] Lee, K.-J., Cho, B.-H., Cho, W.-Y., Kang, S., Choi, B.-G., Oh, H.-R., Lee, C.-S., Kim, H.-J., Park, J.-M., Wang, Q., Park, M.-H., Ro, Y.-H., Choi, J.-Y., Kim, K., Koim, Y.-R., Shing, W.-C., Lim, K.-W., Cho, H.-K., Choi, C.-H., Chung, W.-R., Kim, D.-E., Yu, K.-S., Jeong, G.-T., Jeong, H.-S., Kwak, C.-K., Kim, C.-H., and Kim, K. (2007). A 90 nm 1.8 V 512 Mb diode-switch PRAM with 266 MB/s read throughput. Paper presented at the International Solid-State Circuits Conference, San Francisco, CA.

[14] Villa, C., Mills, D., Barkley, G., Giduturi, H., Schippers, S., and Vimercati, D. (2010). A 45 nm 1 Gb 1.8 V phase-change memory. Paper presented at the International Solid-State Circuits Conference, San Francisco, CA.

[15] Lai, S. and Lowrey, T. (2001). OUM—A 180 nm nonvolatile memory cell element technology for stand alone and embedded applications. Paper presented at the International Electron Device Meeting, Washington, DC.

[16] Yonezawa, F., Nose, S., and Sakamoto, S. (1987). Computer simulations of preparation and annealing of amorphous solids. *Journal of Non-Crystalline Solids*, 97–98: 373–378.

[17] Yeo, E. G., Shi, L. P., Zhao, R., and Chong, T. C. (2006). Investigation on ultra-high density and high speed non-volatile phase change random access memory (PCRAM) by material engineering. MRS Symposium Proceedings, 918, Chalcogenide Alloys for Reconfigurable Electronics, San Francisco, CA.

[18] van Pieterson, L., Lankhorst, M. H. R., van Schijndel, M., Kuiper, A. E. T., and Roosen, J. H. J. (2005). Phase-change recording materials with a growth-dominated crystallization mechanism: A materials overview. *Journal of Applied Physics*, 97(8): 83520–83527.

[19] Cho, S. L., Horii, H., Park, J. H., Yi, J. H., Kuh, B. J., Ha, Y. H., Park, S. O., Kim, H. S., Chung, U. I., and Moon, J. T. (2004). A novel cell technology for phase change RAM. Paper presented at the European Phase Change and Ovonics Symposium.

[20] Lai, Y., Qiao, B., Feng, J., Ling, Y., Lai, L., Lin, Y., Tang, T., Cai, B., and Chen, B. (2005). Nitrogen-doped Ge2Sb2Te5 films for nonvolatile memory. *Journal of Electronic Materials*, 34(2): 176–181.

[21] Ling, Y., Lin, Y., Qiao, B., Lai, Y., Feng, J., Tang, T., Cai, B., and Chen, B. (2006). Effects of Si doping on phase transition of $Ge_2Sb_2Te_5$ films by *in situ* resistance measurements. *Japanese Journal of Applied Physics*, 45(12): L349–L351.

[22] Chen, Y. C., Rettner, C. T., Raoux, S., Burr, G. W., Chen, S. H., Shelby, R. M., Salinga, M., Risk, W. P., Happ, T. D., McClelland, G. M., Breitwisch, M., Schrott, A., Philipp, J. B., Lee, M. H., Cheek, R., Nirschl, T., Lamorey, M., Chen, C. F., Joseph, E., Zaidi, S., Yee, B., Lung, H. L., Bergmann, R., and Lam, C. (2006). Ultra-thin phase-change bridge memory device using GeSb. Paper presented at the International Electron Device Meeting, December.

[23] Happ, T. D., Breitwisch, M., Schrott, A., Philipp, J. B., Lee, M. H., Cheek, R., Nirschl, T., Lamorey, M., Ho, C. H., Chen, S. H., Chen, C. F., Joseph, E., Zaidi, S., Burr, G. W., Yee, B., Chen, Y. C., Raoux, S., Lung, H. L., Bergmann, R., and Lam, C. (2006). Novel one-mask self-heating pillar phase change memory. Paper presented at the Symposium on VLSI Technology, October.

[24] Breitwisch, M., Nirschl, T., Chen, C. F., Zhu, Y., Lee, M. H., Lamorey, M., Burr, G. W., Joseph, E., Schrott, A., Philipp, J. B., Cheek, R., Happ, T. D., Chen, S. H., Zaidi, S., Flaitz, P., Bruley, J., Dasaka, R., Rajendran, B., Rossnagel, S., Yang, M., Chen, Y. C., Bergmann, R., Lung, H. L., and Lam, C. (2007). Novel lithography-independent pore phase change memory. Paper presented at the Symposium on VLSI Technology, June.

[25] Raoux, S., Burr, G. W., Breitwisch, M. J., Rettner, C. T., Chen, Y.-C., Shelby, R. M., Salinga, M., Krebs, D., Chen, S.-H., Lung, H.-L., and Lam, C. H. (2008). Phase-change random access memory: A scalable technology. *IBM Journal of Research and Development*, 52(4–5): 465–479.

[26] Castro, D. T., Guox, L., Hurkx, G. A. M., Attenborough, K., Delhougne, R., Lisoni, J., Jedema, F. J., 't Zandt, M. A. A., Wolters, R. A. M., Gravesteijn, D. J., Verheijen, M. A., Kaiser, M., Weemaes, R. G. R., and Wouters, D. J. (2007). Evidence of the thermo-electric Thomson effect and influence on the program conditions and cell optimization in phase-change memory cells. Paper presented at the International Electron Device Meeting, December.

[27] Cho, W. Y., Cho, B.-H., Choi, B.-G., Oh, H.-R., Kang, S., Kim, K.-S., Kim, K.-H., Kim, D.-E., Kwak, C.-K., Byun, H.-G., Hwang, Y., Ahn, S., Koh, G.-H., Jeong, G., Jeong, H., and Kim, K. (2005). A 0.18-mm 3.0-V 64-Mb nonvolatile phase-transition random access memory (PRAM). *IEEE Journal of Solid-State Circuits*, 40(1): 293–300.

[28] Lee, K.-J., Cho, B.-H., Cho, W.-Y., Kang, S., Choi, B.-G., Oh, H.-R., Lee, C.-S., Kim, H.-J., Park, J.-M., Wang, Q., Park, M.-H., Ro, Y.-H., Choi, J.-Y., Kim, K.-S., Kim, Y.-R., Shin, I.-C., Lim, K.-W., Cho, H.-K., Choi, C.-H., Chung, W.-R., Kim,

D.-E., Yu, K.-S., Jeong, G.-T., Jeong, H.-S., Kwak, C.-K., Kim, C.-H., and Kim, K. (2008). A 90 nm 1.8 V 512 Mb diode-switch PRAM with 266 MB/s read throughput. *IEEE Journal of Solid-State Circuits*, 43(1): 150–162.

[29] Zhang, Y., Kim, S., McVittie, J. P., Jagannathan, H., Ratchford, J. B., Chidsey, C. E. D., Nishi, Y., and Wong, H.-S. P. (2007). An integrated phase change memory cell with Ge nanowire diode for cross-point memory. Paper presented at the Symposium on VLSI Technology, June.

[30] Bedeschi, F., Fackenthal, R., Resta, C., Donzè, E. M., Jagasivamani, M., Buda, E. C., Pellizzer, F., Chow, D. W., Cabrini, A., Calvi, G. M. A., Faravelli, R., Fantini, A., Torelli, G., Mills, D., Gastaldi, R., and Casagrande, G. (2009). A bipolar-selected phase change memory featuring multi-level cell storage. *IEEE Journal of Solid-State Circuits*, 44(1): 217–227.

[31] Kang, D.-H., Lee, J.-H., Kong, J. H., Ha, D., Yu, J., Um, C. Y., Park, J. H., Yeung, F., Kim, J. H., Park, W. I., Jeon, Y. J., Lee, M. K., Song, Y. J., Oh, J. H., Jeong, G. T., and Jeong, H. S. (2008). Two-bit cell operation in diode-switch phase change memory cells with 90 nm technology. Paper presented at the Symposium on VLSI Technology, June.

[32] Xu, W. and Zhang, T. (2010). Using time-aware memory sensing to address resistance drift issue in multi-level phase change memory. Paper presented at the 11th International Symposium on Quality Electronic Design (ISQED), March.

[33] On, H.-R., Cho, B.-H., Cho, W. Y., Kang, S., Choi, B.-G., Kim, H.-J., Kim, K.-S., Kim, D.-E., Kwak, C.-K., Byun, H.-G., Jeong, G.-T., Jeong, H.-S., and Kim, K. (2005). Enhanced write performance of a 64 Mb phase-change random access memory. Paper presented at the IEEE International Solid-State Circuits Conference, February.

[34] Nirschl, T., Philipp, J. B., Happ, T. D., Burr, G. W., Rajendran, B., Lee, M.-H., Schrott, A., Yang, M., Breitwisch, M., Chen, C.-F., Joseph, E., Lamorey, M., Chee, R., Chen, S.-H., Zaidi, S., Raoux, S., Chen, Y. C., Zhu, Y., Bergmann, R., Lung, H.-L., and Lam, C. (2007). Write strategies for 2 and 4-bit multi-level phase-change memory. Paper presented at the International Electron Device Meeting, December.

[35] Hanzawa, S., Kitai, N., Osada, K., Kotabe, A., Matsui, Y., Matsuzaki, N., Takaura, N., Moniwa, M., and Kawahara, T. (2007). A 52 KB embedded phase change memory with 416 KB/s write throughput at 100 mA cell write current. Paper presented at the IEEE International Solid-State Circuits Conference, February.

[36] Lacaita, A. L. and Jelmini, D. (2007). Reliability issues and scaling projections for phase change nonvolatile memories. Paper presented at the International Electron Device Meeting, December.

[37] Lee, S., Jeong, J., Lee, T. S., Kim, W. M., and Cheong, B. (2009). A study on the failure mechanism of a phase-change memory in write/erase cycling. *IEEE Electron Device Letters*, 30: 5.

[38] Yoon, S.-M., Choi, K.-J., Lee, N.-Y., Lee, S.-Y., Park, Y.-S., and Yu, B.-G. (2007). Nanoscale observations of the operational failure for phase-change-type nonvolatile memory devices using $Ge_2Sb_2Te_5$ chalcogenide thin films. *Applied Surface Science*, 254(1): 316–320.

[39] Li, L. and Chan, M. (2008). Scaling analysis of phase change memory (PCM) driving devices. Paper presented at the International Conference on Electron Devices and Solid-State Circuits. December.

[40] Shih, Y. H., Wu, J. Y., Rajendran, B., Lee, M. H., Cheek, R., Lamorey, M., Breitwisch, M., Zhu, Y., Lai, E. K., Chen, C. F., Stinzianni, E., Schrott, A., Joseph, E., Dasaka, R., Raoux, S., Lung, H. L., Lam, C. (2008). Mechanisms of retention loss in $Ge_2Sb_2Te_5$-based phase-change memory. Paper presented at the Electron Devices Meeting, 15–17 December.

3

Spin-Transfer Torque RAM

Spin-transfer torque random-access memory (STT-RAM) is a type of magnetic RAM (MRAM) that uses a magnetic device to store data. Traditional MRAM technology, also called *toggle-mode MRAM*, uses a current-induced magnetic field to switch the data stored in the magnetic pillars. Toggle-mode MRAM faces severe scaling issues because the current required to generate the magnetic field increases as the magnetic pillar becomes smaller. STT-RAM, which is based on spin-polarized current-induced magnetic tunneling junction (MTJ) switching, has attracted a lot of attention due to its fast speed, low power consumption, and better scalability.

3.1 Introduction of Spin-Transfer Torque Technology

3.1.1 Spin-Transfer Torque Theory

The theory of spin-transfer torque was first proposed by Berger [1] and Slonczewski [2] a decade ago and further developed by other researchers, such as Stiles and Zangwill [3]. Those studies showed that a spin-polarized current can reverse the magnetization of a ferromagnetic layer by the spin-transfer effect. The mechanism occurs in a sandwich structure as shown in Figure 3.1, which is composed of two ferromagnetic layers separated by an oxide layer as isolation.

The physical mechanism can be briefly described as follows: In the vicinity of the interface between the normal layer and ferromagnetic layers, the interaction between spin waves and itinerant electrons is considerably enhanced. As a result, the Gilbert damping parameter, which characterizes spin dynamics, increases locally. When a DC current crosses this interface, the stimulated emission of spin waves is expected to occur. When the DC current amplitude exceeds a certain critical current density (Jc), the spin damping becomes negative and a spontaneous precession of the magnetization is predicted to arise [4]. This phenomenon was observed in nanometer-sized pillars and exchange-biased spin valves when current density was around 10^8 A/cm^2 [5, 6].

In 2003, Liu et al. [7] observed that the resistance change driven by current densities above 10^6 A/cm^2 in thin tunnel junctions did not depend on the relative magnetization orientation of the ferromagnetic layers. The current-induced

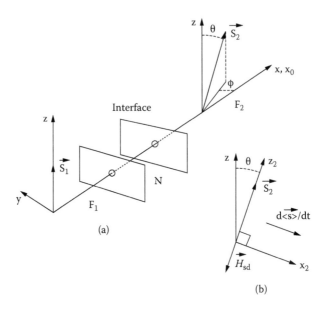

FIGURE 3.1
(a) Coordinate system x, y, z, and polar angles q, j giving the orientation of localized spin $S2$ in layer $F2$. (b) Coordinate system $x2$, $y2$, $z2$ in layer $F2$ with the $z2$ axis parallel to $S2$ and the $x2$ axis in the $(z, S2)$ plane [1].

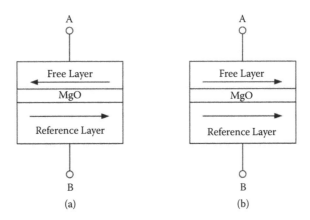

FIGURE 3.2
An MTJ structure: (a) antiparallel (high resistance) and (b) parallel (low resistance) [47].

resistance change is called *current-induced switching* (CIS) and was attributed to electro-migration in nano-constrictions in the insulating barrier [8]. The sandwich structure with two ferromagnetic layers and one oxide barrier layer, for example, MgO, is called a magnetic tunneling junction, shown in Figure 3.2. The spin transfer and current-induced switching effects are likely to coexist in thin MTJs for current densities greater than 10^6 A/cm^2 [9, 10].

3.1.2 Dynamic Behavior of an MTJ

In an MTJ, the magnetization direction of one ferromagnetic layer (called a *reference layer*) is fixed by coupling to a pinned magnetization layer; the magnetization direction of the other ferromagnetic layer (called a *free layer*) can be changed by passing a driving current polarized by a reference layer. As shown in Figure 3.2, the resistance of an MTJ changes corresponding to the magnetization alignment of the two ferromagnetic layers: when their magnetization directions are parallel, the MTJ is in a low-resistance state (R_L); when the magnetization directions of these two layers are antiparallel, the MTJ is in high-resistance state (R_H).

Figure 3.3 illustrates a typical static R-V sweep curve of an MgO-based MTJ [12]. The corresponding electron movements and magnetization rotating of the free layer are shown in Figures 3.4 [13] and 3.5 [14], respectively.

When applying a positive voltage on the free layer of an MTJ, i.e. point A in Figure 3.2, the MTJ resistance could switch from R_H to R_L, which is corresponding to the points 1, 2, and 3 in Figures 3.3 and 3.5. During the whole procedure, injection of a majority of the spin electrons is in the opposite direction, from the reference layer to the free layer. Electrons with a spin polarization parallel to the reference layer can go through it. Electrons with spin polarization antiparallel to the reference layer will be reflected by it. A change in the resistance from point 1 to point 2 is due to the dependence of the tunneling barrier on biased voltage. This mechanism is extremely fast and beyond the normal transient region of STT-RAM operation [19, 27]. The resistance change from point 2 to 3 is due to the magnetization rotating of the free layer under the influence of the spin polarization current. When the current density exceeds the switching current density I_c, the free-layer magnetization is changed to be parallel to the reference layer.

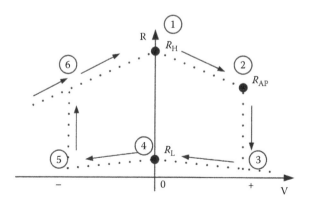

FIGURE 3.3
Illustration of static R-V curve of an MTJ [27].

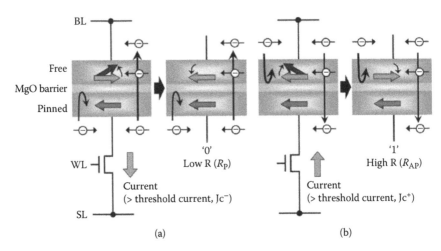

FIGURE 3.4
STT-RAM write operations: (a) '0' write, parallelizing; and (b) '1' write, antiparallelizing [45].

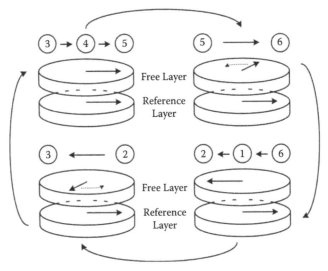

FIGURE 3.5
Dynamic behavior of an MTJ [47].

When applying a positive voltage on point B in Figure 3.2, the MTJ enters the negative applied voltage region shown in Figure 3.3 and switches from R_L to R_H, which corresponds to points 4, 5, and 6. At this time, injection of a majority of the spin electrons is from the free layer to the reference layer. Electrons with antiparallel spin polarizations as the reference layer are reflected by the reference layer, which can influence the free-layer magnetization. Simultaneously, the resistance of the MTJ switches from R_L to R_H continuously. Electrons with the parallel spin polarization as the reference layer can go through the reference layer. Due to the dependence of the tunneling barrier on biased voltage, the resistances of MTJ at point 5 is slightly lower than at point 5. After removing the positive voltage on point B shown in Figure 3.2, the resistance of the MTJ will change from point 6 to point 1. The magnetization rotating of the free layer under the influence of the spin polarization current results in the resistance change from point 5 to point 6, shown in Figure 3.3.

3.1.3 Some Important Parameters of MTJ

3.1.3.1 R_H, R_L, R_{AP}, and Tunneling Magnetoresistance

The resistance states of an MTJ correspond to the magnetization alignment of the two ferromagnetic layers. Usually, R_H refers to the high-resistance state when the magnetization directions of the reference layer and free layer are antiparallel. When the magnetization directions of these two layers are parallel, the MTJ is in a low-resistance state (R_L).

The static *R-V* sweep curve of an MTJ, shown in Figure 3.3, shows that the MTJ resistance relies on the bias voltage. Usually, R_H and R_L are respectively used to represent the high- and low-resistance states of an MTJ when the bias voltage is close to 0. As the bias voltage increases, the real high (or low) resistance of the MTJ decreases because of the tunneling barrier effect [11]. We refer to the antiparallel resistance R_{AP} as the high-resistance state of an MTJ at which the free layer enters the magnetization rotating stage. The change in the MTJ high-resistance state is denoted as $\Delta R_H = R_H - R_{AP}$. Similarly, the parallel resistance R_P is the low-resistance state of an MTJ at which the magnetization direction of the free layer starts rotating from parallel to antiparallel. The change in the MTJ low-resistance state ($\Delta R_L = R_L - R_P$) is so small that it is usually negligible.

The tunneling magnetoresistance (TMR) ratio represents the gap between the high- and low-resistance states of an MTJ. It is defined as

$$TMR = \frac{R_H - R_L}{R_L} \tag{3.1}$$

3.1.3.2 Aspect Ratio

The aspect ratio of an MTJ device is the ratio of its longer dimension to its shorter dimension. To maintain the required thermal stability, a certain aspect ratio of the MTJ device must be kept; that is, 2:1. Such geometry constraint could limit the scalability of the STT-RAM cell area. A magnetic solution, which introduces a surface antiferromagnetic-coupled (AFC) magnetic layer with relatively low Curie temperature [18], can further reduce MTJ shape to a circle.

3.1.3.3 Resistance-Area Product

RA represents the resistance and area product of an MTJ device, which is determined by the material and structure of the MTJ. *RA* is an intrinsic parameter and independent of the MTJ's physical dimensions. For a specific MTJ structure, it changes according to the thickness of the oxide barrier.

The switching current I_C is proportional to the physical area of an MTJ device. Therefore, as technology scales and the MTJ dimensions become smaller, the switching current decreases accordingly. This is why the spin-transfer torque technique has advantages for further scaling.

3.1.3.4 Switching Current I_C

Switching current, the required magnitude of the writing current to switch the MTJ resistance, is an important electrical parameter in STT-RAM design. It is an intrinsic parameter that relies on the MTJ's material and structure. In addition, it is heavily related to the write pulse width, or switching time, which is defined as the length of time that the switching current is applied to the MTJ.

Figure 3.6 demonstrates a pulse width dependence on the switching current for an MTJ with a size of 115 nm × 155 nm [11]. Threshold currents were averaged from 50 cycle measurements on each pulse width. As write pulse width was reduced, both write threshold current and current distributions were increased.

For a relatively long write pulse (10 ns or longer), the required magnitude of switching current agrees well with the theoretical equation [11]:

$$I_C = I_{C0}\left\{1-\left(\frac{kT}{E}\right)\ln\left(\frac{\tau}{\tau_0}\right)\right\} \tag{3.2}$$

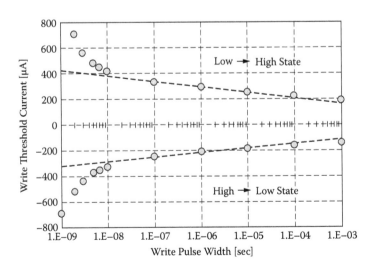

FIGURE 3.6
Switching currents on different pulse widths, MTJ size of 115 × 155 nm. Threshold currents were averaged from 50 cycle measurements on each pulse width. As the write pulse width was reduced, both write threshold current and current distributions were increased [11].

Here τ is the write pulse width; I_C is the critical switching current, which is the minimum required magnitude of switching current to switch the MTJ resistance by τ; I_{C0} is the critical switching current at 0 K; E is the magnetization stability energy barrier; $τ_0$ is the inverse of the attempt frequency or the write pulse width at 0 K; k is the Boltzmann constant; and T is the operating temperature.

When the write pulse width is shorter than 10 ns, short-time magnetization dynamics dominate. Reducing the write pulse width will increase the critical switching current rapidly [12–14]. The critical switching current in the whole time range, from the long-time thermal reversal region to the short-time dynamic region, can be modeled by the stochastic Landau-Lifshitz-Gilbert dynamic equation with a spin-torque term [15].

3.1.3.5 MTJ Switching Probability P_{sw}

MTJ switching probability is an important factor in understanding the switching behavior of an MTJ. From a theoretical point of view, the magnetization direction switching of the free layer depends on magnetization stability energy barrier height. Hence, the MTJ switching probability can be expressed as a function of the magnetic memorizing energy Δ as follows [16, 17]:

$$P_{sw} = 1 - \exp\left\{\frac{t}{τ_0}\exp\left[-Δ_0\left(1 - \frac{1}{I_{C0}}\right)\right]\right\} \qquad (3.3)$$

Here, $Δ_0$ represents the magnetic memorizing energy without any current and magnetic field, t is the pulse width, I_{C0} is the critical switching current at 0 K and $τ_0$ is the relaxation time at 0 K. Figure 3.7 shows an experimental result of the switching probability for an MTJ cell with a size of 100 nm

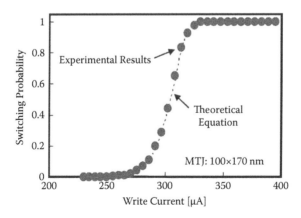

FIGURE 3.7
Switching probability versus write current [11].

× 170 nm [11]. The measurement result is consistent with the theoretical calculation based on the above equation.

3.2 Spin-Transfer Torque Random Access Memory

3.2.1 Basic Structure

The most popular design of STT-RAM is a one transistor-one MTJ (or 1T-1J) structure. The simple structure uses an MTJ as the storage element and an N-channel metal-oxide semiconductor field-effect transistor (NMOS) as the selection device. Diagrams of the circuit, cross section, and layout are shown in Figure 3.8.

In an STT-RAM cell, the terminal direct to the source of the NMOS transistor is called a *source line* (SL). Usually the lowest metal layer in a complementary

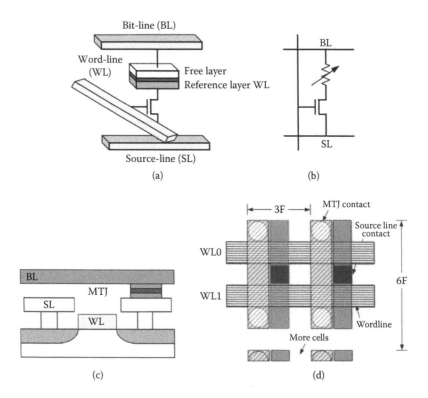

FIGURE 3.8
1T-1J STT-RAM structure: (a) device view [27]; (b) circuit diagram [27]; (c) cross section [45]; and (d) layout [47].

metal-oxide semiconductor (CMOS) process, that is, metal 1, is used for SL wiring in an STT-RAM layout. A bit line (BL), which is the terminal to the MTJ free layer, uses a back-end-of-line (BEOL) process. The pin layer of the MTJ and the drain of the NMOS transistor are connected together. The terminal connected to the gate of an NMOS transistor is called a *word line* (WL). Polysilicon or metals could be used for WL wiring. For simplicity, an MTJ is usually represented as a variable resistor in circuit schematics.

A diagram of an STT-RAM cell layout is shown in Figure 3.8(d). Along the *x*-axis, two metal tracks are reserved for the BL and SL, respectively, in each cell. Shallow trench isolation (STI) is needed to separate two adjacent transistors. Two adjacent STT-RAM cells in the same column can share the source of NMOS transistors. In an ideal situation, the width of metal/diffusion/STI uses the minimal feature size (F), an STT-RAM design achieves the smallest area of $9F^2$.

3.2.2 Write and Read Mechanism

As discussed in Section 3.1.2, the switching of MTJ resistance change is accomplished by controlling the direction of the current through the MTJ. Accordingly, writing different data to an STT-RAM cell can be controlled by providing the different current polarity. In '0' writing, the current flows from the BL to the SL—that is, from the free layer to the reference layer of the MTJ, as shown in Figure 3.8(a)—whereas '1' writing operations can be achieved with a contrary current flow; that is, from the SL to the BL.

Figure 3.9 illustrates a common read-out scheme of STT-RAM. By applying a read voltage (current) to the selected memory cell the generated current (voltage) on the bit line can be compared to a reference signal in the sense

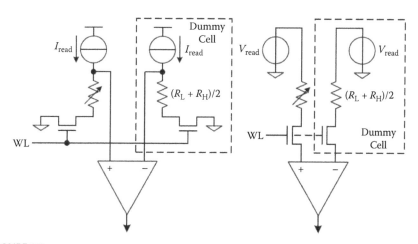

FIGURE 3.9
Conventional read-out scheme of an STT-RAM [23].

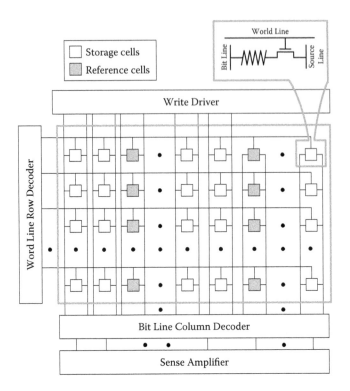

FIGURE 3.10
STT-RAM array architecture.

amplifier. If the generated current (voltage) is higher than the reference, the data storage device in the memory cell is in the low- (high-) resistance state. The reference signal is normally generated by applying the same read voltage (current) on the dummy cell, whose resistance is ideally $(R_L + R_H)/2$.

3.3 STT-RAM Architecture

As we can see from Figure 3.10, the architecture of an STT-RAM array is similar to that of static RAM (SRAM). This is because STT-RAM and SRAM cells have exactly the same terminals; that is, WL, BL, and SL ($\overline{BL}$ in SRAM).

However, there are still some differences between STT-RAM and SRAM architectures. For example, precharging BL and $\overline{BL}$ is required before reading operations in SRAM, which is not necessary in STT-RAM. Moreover, some STT-RAM designs use the extra reference cells, which can be shared by storage cells, in read operations. The smaller cell size of STT-RAM makes its layout even more difficult; in particular, the layout of peripheral circuitry needs to become shorter and/or thinner compared to that in SRAM design.

3.4 STT-RAM Modeling

3.4.1 Dynamic MTJ Modeling

Based on the spin-transfer torque model proposed by Berger [1] and Zhang et al. [4], a dynamic MTJ model for STT-RAM design can be generated. In the model, the dynamics of the free layer magnetization can be simulated using the Landau-Lifshiltz-Gilbert (LLG) equation with a spin-polarized torque term:

$$\frac{d\vec{m}}{dt} = \alpha\vec{m}\times(\vec{m}\times\vec{h}_{\text{eff}}) - \vec{m}\times(\vec{h}_{\text{eff}} + \beta\vec{m}\times\vec{p}) \tag{3.4}$$

On the right side of Eq. (3.4), the first term shows the effect of spin damping, where $\vec{m}$ is the normalized magnetization, $\vec{h}_{\text{eff}} = \vec{H}_{\text{eff}}/M_{\text{s}}$ is the normalized magnetic field (by shape and anisotropy, etc.), and α is the damping parameter. The second term on the right side of Equation (3.4) is the spin torque term. $\vec{p}$ is a unit vector pointing in the spin polarization direction and $\beta = \eta hJ/2eM_{\text{s}}$ is the normalized spin torque polarization magnitude, where β is a function of spin polarization efficiency η, magnetic film saturation M_{s}, thickness d, surface S, and the switching current density J through the MTJ [1, 4]. Notice that here the spin torque term only includes an adiabatic term. Another term proportional to $\vec{m}\times\vec{p}$ can be added to Equation (3.4) for nonadiabatic spin-torque effects.

The model considers the transient response of the MTJ with time-varying electrical inputs; that is, current density through the MTJ. For example, during the procedure in which the MTJ resistance changes from R_{AP} to R_{L} or vice versa, the amplitude of the current density J could be a time-varying quantity that is affected by the dynamics of the MTJ and other circuit components, such as the NMOS transistor. This model can handle the interaction between the MTJ and circuit components.

3.4.2 Cell Modeling

Figure 3.11 shows a simplified schematic of the dynamic behavior of an STT-RAM cell: the MTJ is represented by a variable resistor; the NMOS transistor is modeled as a voltage (V_{GS}, V_{DS}, and V_{SB}) controlled current source (VCCS) I_{T}, which can be precharacterized by Simulation Program for Integrated Circuits Emphasis (SPICE) simulation, and a set of parasitic capacitors. Here C_{GD}, C_{GS}, C_{DB}, C_{SB}, and C_{GB} denote the capacitances between the gate and drain, gate and source, drain and body, source and body, and gate and body, respectively. Usually C_{GB} can be ignored [10].

The dynamic electrical behavior of a 1T-1J STT-RAM cell can be expressed by:

$$I_M + I_{GD} = I_{DB} + I_T$$

$$I_{GD} = C_{GD} \cdot \left(-\frac{dV}{dt} \right)$$

$$I_{DB} = C_{DB} \cdot \left(\frac{dV}{dt} \right)$$

(3.5)

$$I_M \cdot R(t) = V_{DD} - V \quad \text{(current from BL to SL)} \quad \text{or}$$

$$I_M \cdot R(t) = -V \quad \text{(current from SL to BL)}$$

Here, I_M, I_{GD}, I_{DB}, and I_T are the currents through the MTJ, C_{GD}, C_{DB}, and NMOS transistor, respectively. V is the voltage at the point between the MTJ and the NMOS transistor, and $R(t)$ is the time-varying MTJ resistance.

By combining Eqs. (3.4) and (3.5), a dynamic STT-RAM cell model that can take into account both the magnetic response of MTJ and the electrical response of the NMOS transistor is achieved.

Compared to the conventional static STT-RAM cell model, the above dynamic STT-RAM cell model is much closer to the realistic physical mechanism of an STT-RAM write operation.

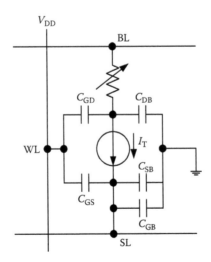

FIGURE 3.11
Simplified schematic to model the dynamic behavior of an STT-RAM cell [47].

3.5 Design Challenges

3.5.1 Sources of Process Variations

Like all the other memory designs at the nanometer scale, STT-RAM faces tremendous challenges due to process variations, which come from both magnetic devices and CMOS technology.

3.5.1.1 CMOS Process Variations

CMOS process variations—for example, doping and geometry variations—can affect the driving ability and the equivalent resistance of transistors and can therefore impact the STT-RAM design. For instance, the NMOS selection transistor in an STT-RAM cell works in the linear region during '1' writing operations. A small deviation of the NMOS device parameter or a small fluctuation of gate-to-source voltage could result in a big difference in the switching current through the MTJ.

3.5.1.2 MTJ Resistance Variations

Because an MTJ is the key component for information storage, the stability of an STT MRAM bit cell primarily depends on the characteristics of the MTJ device. It has been experimentally demonstrated that the sources of variations in process parameters of MTJ mainly arise from variations in tunneling oxide thickness (τ) and cross-sectional area (A) [34]; the impact on MTJ resistance in parallel and antiparallel configurations under different bias voltages (V_{MTJ}) are shown in Figure 3.12. Due to quantum mechanical tunneling, the thickness of the tunneling film (τ) affects R_{AP} and R_P in an exponential manner. Hence, a small variation in τ can lead to large spread in R_{AP} and R_P. Cross-sectional area (A) variation affects both electrical static and magnetic dynamic switching characteristics of the MTJ. In static states, the resistance is inversely proportional to A. Due to the time-dependent stochastic magnetic switching process, variations in A introduce a linear shift in switching threshold current, as shown in Figure 3.13.

3.5.1.3 Switching Current Variations

As introduced above, the required switching current is an intrinsic parameter of MTJ, which is determined by the MTJ material and structure and is related to the switching time. In STT-RAM design, the switching current is controlled by the peripheral circuitry, such as address transition detection (ATD). For a given switching time, the switching current through an MTJ is impacted by many factors, such as MTJ resistance, NMOS selection

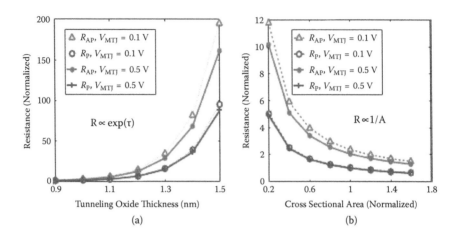

FIGURE 3.12
Impact of (a) tunneling oxide thickness (τ) and (b) cross-sectional area (A) in MTJ resistance (R) in parallel and antiparallel configuration under different bias voltage (V_{MTJ}). The values are normalized to the parallel case when $\tau = 1.0$ nm, $V_{MTJ} = 0.1$ V [34].

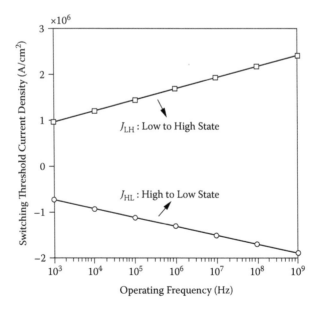

FIGURE 3.13
Frequency dependency of switching threshold current density of an MTJ. J_{HL} and J_{LH} are the switching threshold current density from high to low and from low to high resistive states, respectively [34].

transistor variations, power supply noise, etc. Therefore, the switching current in a real implementation could vary from chip to chip or even from cell to cell due to process variations. Usually, at the design stage, a long switching time should be set up to guarantee that the MTJ state can be changed even in the worst-case scenario.

3.5.2 Static Failures in STT-RAM

In principle, the cell failure in STT-RAM can be divided into either (1) read failure, which occurs due to destructive read or the wrong sensing results at read out circuit, or (2) write failure due to unsuccessful write operations [34].

During read operations, a sense amplifier is used to compare the voltage/current through a BL with the reference value. Due to process variations, improper design, noise, etc., a wrong decision could be made by the sense amplifier. We refer to this type of read failure as a *decision failure*, which is illustrated in Figure 3.14. Another read failure mechanism arises from the read current disturbance; that is, a cell could flip its value during read operations. This is referred to as a *disturbance failure*. The corresponding current distribution is illustrated in Figure 3.15. Due to the hysteretic nature of an MTJ, there is only one direction disturbance; that is, "1-to-0 disturbance" when in the parallel direction reading or "0-to-1 disturbance" when in the antiparallel direction reading. Estimation of read failure probability can be found in the literature [4, 34, 43, 46].

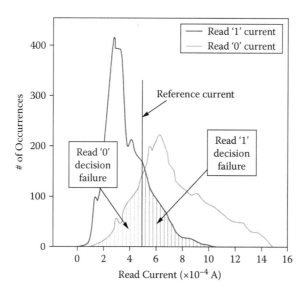

FIGURE 3.14
Current distributions to illustrate decision failure [46].

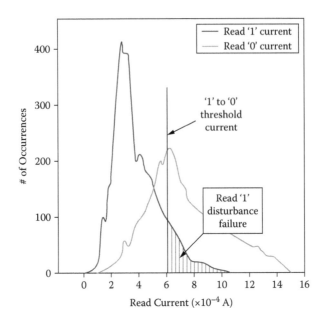

FIGURE 3.15
Current distributions to illustrate disturbance failure [46].

Unsuccessful write occurs when the writing current is lower than the switching threshold current or the write switching time is not long enough. The current distribution in Figure 3.16 can be used to illustrate the write failures in STT-RAM. The estimation of write failure probability is discussed in J. Li et al. [46].

3.6 Design Techniques of STT-RAM

3.6.1 STT-RAM Cell

3.6.1.1 Other STT-RAM Cell Designs

In addition to the most popular STT-RAM cell design—the one transistor-one-MTJ (1T-1J) structure shown in Figure 3.8—many other cell designs, such as two-transistor-one-MTJ (2T-1MTJ) [34] and thermal-assisted programming [37, 39, 40, 51] have been investigated.

2T-1MTJ

In the conventional 1T-1J cell, it is difficult to size the NMOS transistor to achieve both read and write robustness to process variation. Therefore, J. Li et al. proposed a 2T-1MTJ design technique that can compensate cell stability

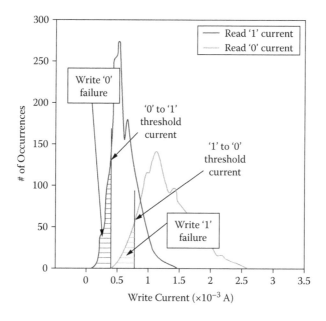

FIGURE 3.16
Current distributions to illustrate write failure [46].

in read operations and improve MTJ switching probability in write operations by scarifying memory density [34]. The corresponding cell schematic is shown in Figure 3.17. Compared to the conventional 1T-1J configuration shown in Figure 3.8, a 2T-1MTJ cell consists of two NMOS transistors—a read-NMOS and write-NMOS—in parallel with independent gate control. In a read operation, only the read-NMOS is turned on to achieve optimal read immunity to disturbance; during a write operation, the read-NMOS and write-NMOS are simultaneously turned on to provide a relatively large current for better write stability.

This technique effectively decouples the conflict between read stability and write stability, resulting in considerably improved robustness. However, cell area increases due to the two transistors. The layout of a 2T-1MTJ structure is shown in Figure 3.18.

Thermal-Assisted Cell Programming Scheme

Thermal-assisted programming schemes were first proposed for conventional MRAM [8, 35] to reduce the critical switching current. The only difference between the standard STT-RAM and thermal-assisted STT-RAM is the memory bit stack design. In thermal-assisted STT-RAM, the free layer of the basic MTJ stack is coupled with an antiferromagnetic (AF) layer. At room temperature, the free layer magnetization is stabilized by the exchange bias arising from the atomic contact with the AF layer. When programming a thermal-assisted memory bit, an electric current (heating current)

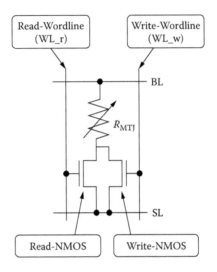

FIGURE 3.17
2T-1MTJ structure [34].

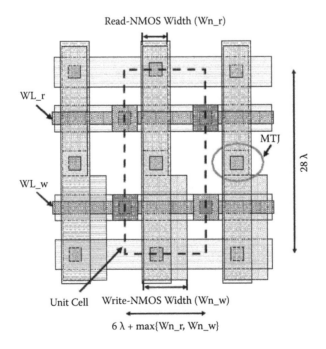

FIGURE 3.18
Layout of 2T-1MTJ structure [34].

is driven through the MTJ stack and heats it up. After the temperature on the MTJ rises to the blocking temperature T_B, at which the exchange bias effect vanishes [36], the free layer is free to rotate. The spin torque of the electric current then switches the free layer to a parallel state or an antiparallel state, depending upon its polarity. After the heating current is removed, the MTJ temperature drops below T_B. The exchange bias on the free layer then appears again and stabilizes the free layer magnetization in the new direction. In the thermal-assisted programming process, the critical switching current is not determined by the spin torque strength required to switch the free layer magnetization but rather by the temperature that the current can heat the MTJ up to. Due to reliability issues, the blocking temperature T_B of the exchange bias on the free layer must be beyond the memory operation temperature range. A reasonable choice of T_B, for example, is 150°C [37].

It has been well demonstrated in Daughton and Pohm [38] and Prejbeanu et al. [39] that a thermal-assist scheme can substantially reduce the write current in conventional MRAM. Similarly, STT-RAM can benefit from thermal assistance for better write ability and gain advantages in the manufacturing process [40]. A typical MTJ stack of thermal-assisted STT-RAM is shown in Figure 3.19 [51]. The MTJ stack forms a multilayer that is composed of, from bottom to top, a seed layer (3 nm)/PtMn (20 nm)/CoFe (3 nm)/Ru (1 nm)/ CoFeB (3 nm)/MgO/CoFeB (2 nm)/IrMn (5 nm)/cap layer (3 nm). In addition, the MTJ pillar is encapsulated by an Si_3N_4 spacer for oxidation protection.

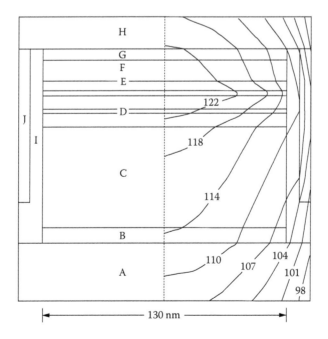

FIGURE 3.19
Thermal assist MTJ design [51].

The whole MTJ element is surrounded with dielectric. Both the spacer and dielectric are thermally resistive materials.

3.6.1.2 STT-RAM Cell Scalability

Theoretically, the minimum area of a 1T-1J STT-RAM cell is $9F^2$, where F represents the technology feature size. As shown in Figure 3.8, we assume that the minimum pitch of each element is $1F$. Moreover, to achieve this $9F^2$ cell area, the critical switching current must be further reduced so that it can be provided by the NMOS transistor with a channel width of $2F$. And the shape of the MTJ must be further scaled to a circle with $1F$ diameter; for example, 22 nm × 22 nm at the 22-nm technology node.

   Tables 3.1 and 3.2 list the MTJ and corresponding NMOS transistor parameters, respectively, when scaling down STT-RAM design from 90- to 22-nm technology. The first three rows only consider direct scaling, which shrinks the MTJ geometrical dimensions but keeps the same material stack.

   The MTJ stack with perpendicular anisotropy can further reduce the critical switching current at the 22-nm technology node. However, to maintain the required thermal stability, certain physical dimensions of the MTJ must be maintained; that is, the 2:1 ratio in the x and y directions (or 45 nm × 22 nm at the 22-nm technology node). We refer to this MJT stack with perpendicular anistropy as magnetic solution 1, which is shown in row "22^1" in Tables 3.1 and 3.2. The physical realization of the MJT with perpendicular anisotropy and the corresponding magnetic theory are beyond the scope of this chapter. We refer the reader to Sun et al. [49] and Kiselev et al. [50] for more detailed discussion.

   As discussed previously, theoretically, the minimum area of a 1T-1J STT-RAM cell that can be achieved is $9F^2$, as shown in Figure 3.8(d). In order to achieve this cell area, the critical switching current must be further reduced so that it can be provided by the NMOS transistor with a channel width of $2F$. In addition, the shape of the MTJ must be further scaled to a circle with a $1F$ diameter (or 22 nm × 22 nm at the 22-nm technology node). However, an MTJ with a spherical or square shape will lead to a severe thermal stability issue [42]. One solution is to introduce a surface AFC magnetic layer with relatively low Curie temperature (we refer to this solution as magnetic solution 2 in Tables 3.1 and 3.2) [8]. The AFC coupling provides additional surface anisotropy to maintain the thermal stability of the MTJ. During the write operation, the joule heating of the spin-torque current raises the temperature of the AFC surface magnetic layer above the Curie temperature. Consequently, the AFC-induced surface anisotropy disappears. The critical switching current is reduced accordingly. Row "22^2" in both Tables 3.1 and 3.2 gives the key parameters of an STT-RAM cell with an AFC-coupled surface layer. The surface exchange coupling parameter of the AFC-coupled surface layer is 0.1 erg/cm^2. For more details of the AFC coupling structure, we refer our readers to Deac et al. [8].

TABLE 3.1

MTJ Parameters

Tech. (nm)	MTJ Geometry (nm)			I_{drv} (μA)	M_s (emu/cc)	R_H (Ω)	R_{AP} (Ω)	R_L (Ω)	RA_L (Ω μm²)	KV/kT	Technology Requirement
	Length	Width	Thickness								
90	110	60	2.2	300	1,000	1,200	1,000	600	4.5	52	Present
45	90	40	2.3	170	1,000	2,000	1,666	1,000	3.6	52	Direct scale down
220	45	22	3.2	100	1,200	3,000	2,500	1,500	1.5	50	Direct scale down
221	45	22	3.2	75	1,200	3,000	2,500	1,500	1.5	50	Magnetic solution 1
222	22	22	2.0	54	1,200	3,000	2,500	1,500	1.5	50	Magnetic solution 2

The worst-case scenario (six-sigma) switching time of MTJ is fixed at 15 ns for all technology nodes.

TABLE 3.2

Parameters of NMOS Transistor

Tech (nm)	Channel Width (nm)	V_{DD} (V)	Area (F^2)	C_{DB} (fp)	C_{GB} (fp)
90	456	1.2	20.6	0.246	0.127
45	273	1.0	21.2	0.203	0.105
220	106	1.0	17.4	0.111	0.047
221	75	1.0	13.2	0.086	0.032
222	45	1.0	9.1	0.062	0.017

3.6.1.3 Variation Control of STT-RAM Bit Cell

In addition to improving the process uniformity and consistency, enhancing the resolution of lithography technology and minimizing distortion are other ways to reduce the impact of process variations. Optical proximity correction (OPC) is one of the most powerful techniques that predistorts the mask data to achieve the desired pattern on the wafer [24]. An OPC algorithm can invert the transfer fuction based on a highly phenomenological process model that incorporates lumped optics, resist, wafer stack, and mask effects and therefore modifies the mask accordingly.

Dummy fill is a must in memory design in order to ensure layout uniformity and satisfy the prescribed density criteria. Some design rules need to be met for efficient dummy fill. First, the dummy cell should be realized and included during resistance-capacitance (RC) parasitic estimation; second, functional cell and macro characterizations must be a priori compatible with the later insertion of dummy fill; and third, dummy fill should be consistent with the design hierarchy to reduce the complexity of design verification. In STT-RAM design, dummy fill must be inserted at all of the metal layers as well as the magnetic layers.

3.6.2 STT-RAM Circuit Design—Self-Reference Scheme

Due to process variations, decision failure is a severe issue in STT-RAM design. Consequently, many sensing schemes have been proposed to alleviate the situation [23] [44]. In this section, we will briefly introduce the existing sensing schemes of STT-RAM and discuss the impacts of device variations on these sensing schemes.

3.6.2.1 Conventional Sensing Scheme

Figure 3.20 illustrates a conceptual schematic of a conventional sensing scheme (CSS), which was briefly introduced in Section 3.2.2. Here we provide a further discussion on the operating mechanism in order to understand why this scheme is intolerable to process variation.

In a CSS of STT-RAM, a read current I_R is applied to an STT-RAM cell and generates the BL voltage V_{BL}. The value of a memory bit can be read out by

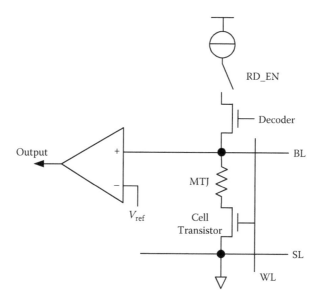

FIGURE 3.20
Conventional sensing scheme.

comparing V_{BL} with a reference voltage V_{REF}, which is generated by applying the same read current on the dummy cell as mentioned in Section 3.2.2. V_{REF} should satisfy

$$V_{BL,L} = I_R \cdot \left(R_L + R_{TR,L} \right) < V_{REF} < V_{BL,H} = I_R \cdot \left(R_H + R_{TR,H} \right) \qquad (3.6)$$

where $V_{BL,L}$ and $V_{BL,H}$ respectively stand for the V_{BL} when the MTJ is in the low- and high-resistance states. R_{TR} represents the transistor resistance and we assume that it remains unchanged during a read operation because the transistor's biasing condition keeps constant. The meaning of Eq. (3.6) is that the sensing voltage in a high- (low-) resistance state of STT-RAM must be higher (lower) than the reference voltage.

Due to process variations, the MTJ resistance varies from bit to bit. If a V_{REF} is shared by multiple bits, the overlap between the distributions of the two MTJ resistance states will cause false sensing. Therefore, the conventional sensing scheme works properly only when

$$\max\left(R_L + R_{TR} \right) < \min\left(R_H + R_{TR} \right) \qquad (3.7)$$

The increased process variations in the scaled technology can incur a large standard deviation of MTJ high and low state resistances. Equation (3.7) might not be satisfied when all of the memory cells are considered, which consequently results in read failures. Although the read failures can be fixed by either the redundancy or error correction code (ECC), the poor robustness of a CSS severely limits the chip yield in STT-RAM design.

3.6.2.2 Conventional Self-Reference Sensing Scheme

The motivation of a conventional self-reference sensing scheme (CSR) is to directly compare the BL voltage generated by the original data stored in an MTJ with the BL voltage generated by reference data stored in the same MTJ. Because the reference signal is generated from the same memory bit, the MTJ resistance variation incurred by the bit-to-bit variations of MTJs is excluded from the sensing operation [7, 8]. Figure 3.21 illustrates a conceptual schematic of the conventional self-reference sensing scheme. The read procedure is as follows:

Read the original data: Apply a read current I_{R1} to generate BL voltage V_{BL1}, which is stored in a capacitor C1. V_{BL1} can be either $V_{BL,L1}$ or $V_{BL,H1}$ when the MTJ is in a low- or high-resistance state, respectively.

Write '0': Value '0' is written into the same MTJ.

Read '0': Another read current I_{R2} is applied and generates BL voltage V_{BL2}, which is stored in capacitor C2. Here I_{R2} is moderately greater than I_{R1} in order to satisfy $V_{BL,L1} < V_{BL2} < V_{BL,H1}$ or:

$$I_{R1} \cdot \left(R_{L1} + R_{TR1} \right) < I_{R2} \cdot \left(R_{L2} + R_{TR2} \right) < I_{R1} \cdot \left(R_{H1} + R_{TR1} \right) \qquad (3.8)$$

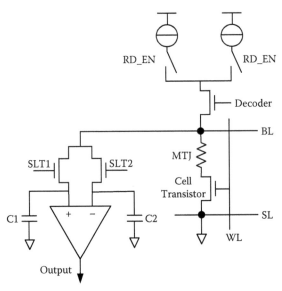

FIGURE 3.21
Conventional self-reference sensing scheme.

The original value of the STT-RAM bit can be read out by comparing V_{BL1} and V_{BL2}.

Write back: Write the original data back to the memory bit.

Here R_{L1} and R_{L2} are the resistances of the MTJ in the low-resistance state under read currents I_{R1} and I_{R2}, respectively. R_{H1} is the resistance of the MTJ in the high-resistance state under read current I_{R1}. R_{TR1} and R_{TR2} are the resistances of the NMOS transistor under currents I_{R1} and I_{R2}, respectively. Usually one set of CSR circuitry is shared by multiple bits.

If we assume the read current ratio of $\alpha = I_{R2}/I_{R1}$ can be flexibly adjusted in the postsilicon stage, CSR works properly for multiple bits only when a valid selection of α exists [13] or:

$$\max\left(\frac{R_{L1}+R_{TR1}}{R_{L2}+R_{TR2}}\right) < \min\left(\frac{R_{H1}+R_{TR1}}{R_{L2}+R_{TR2}}\right) \tag{3.9}$$

Because CSR compares the data within the same MTJ, the impact of process variations can be minimized. However, the two write operations of CSR introduce a long latency period and a large power overhead.

3.6.2.3 Nondestructive Self-Reference Sensing Scheme

A nondestructive self-reference sensing scheme (NDSR) was proposed [23] based on the unique characteristic of MgO-based MTJs: the current roll-off slope of the high-resistance state is much steeper than that of the low-resistance state, as shown in Figure 3.3. NDSR includes only two read steps without overwriting the original value in STT-RAM bit. Figure 3.22 illustrates the design of an NDRS [38]. The read procedure is as follows:

First read: Apply a read current I_{R1} to generate BL voltage V_{BL1}, which is stored in a capacitor C1. V_{BL1} can be either $V_{BL,L1}$ or $V_{BL,H1}$ when the MTJ is in a low- or high-resistance state, respectively.

Second read: Another read current I_{R2} ($I_{R2} > I_{R1}$) is applied and generates BL voltage V_{BL2}. V_{BL2} is translated into the output voltage V_{BL2O} by a voltage divider with a voltage ratio of $\beta = V_{BL,L2O}/V_{BL2}$. The selection of β and read current ratio $\alpha = I_{R2}/I_{R1}$ ensures:

$$V_{BL,L1} = I_{R1}\cdot(R_{L1}+R_{TR1}) < V_{BL,L2O} = \alpha\cdot I_{R2}\cdot(R_{L2}+R_{TR2})$$

$$V_{BL,H2O} = \alpha\cdot I_{R2}\cdot(R_{H2}+R_{TR2}) < V_{BL,H1} = I_{R1}\cdot(R_{H1}+R_{TR1}) \tag{3.10}$$

Here $V_{BL,L1}$ and $V_{BL,H1}$ are the BL voltages when the MTJ resistance equals R_{L1} or R_{H1} at I_{R1}, respectively. $V_{BL,L2O}$ and $V_{BL,H2O}$ are the output of the voltage divider when the MTJ resistance equals R_{L2} or R_{H2} at I_{R2}, respectively.

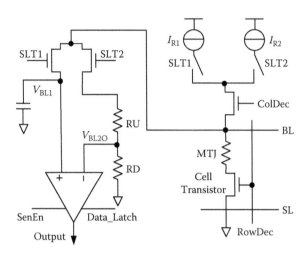

FIGURE 3.22
Nondestructive self-reference sensing scheme [44].

Sensing: The memory bit is read out by comparing V_{BL1} and V_{BL2O}.
For multiple bits, a valid selection exists if:

$$\max\left(\frac{R_{L1} + R_{TR1}}{R_{L2} + R_{TR2}}\right) < \min\left(\frac{R_{H1} + R_{TR1}}{R_{H2} + R_{TR2}}\right) \quad (3.11)$$

The left side of the working conditions of CSR and NDSR both equal $\left(R_{L1} + R_{TR1}\right)/\left(R_{L2} + R_{TR2}\right)$, which is close to 1 because $R_{L1} \approx R_{L2}$ and $R_{TR1} \approx R_{TR2}$. The right side $\left(R_{H1} + R_{TR1}\right)/\left(R_{H2} + R_{TR2}\right) < \left(R_{H1} + R_{TR1}\right)/\left(R_{L2} + R_{TR2}\right)$ because $R_{H2} > R_{L2}$. This explains why NDSR has a relatively smaller process variation tolerance margin than CSR.

Figure 3.23 shows the timing diagram of our nondestructive self-reference scheme. Different read currents are applied to the STT-RAM cell by turning on STL1 or STL2, respectively. Sensing is triggered by the signal "SenEn" and the output of the sense amplifier is captured and stored in a latch by enabling the signal "Data_latch." Figure 3.24 shows the simulation results of our nondestructive self-reference scheme. The circuitry is implemented with Taiwan Semiconductor Manufacturing Company Limited (TSMC) 0.13 mm technology. The whole read operation can complete in about 15 ns.

An NDSR has relatively tighter constraints on the device variations than that of a conventional "destructive" self-reference scheme. However, this technique enables much faster read speed by eliminating two write steps (erase and write back) and shortening the second read step. The reliability of STT-RAM is improved by maintaining the nonvolatility. The sense margin and the robustness of an NDSR can be improved by increasing the maximum allowable read current I_{max} (I_{R2}).

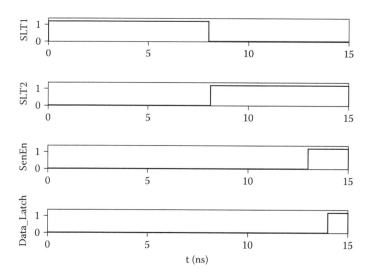

FIGURE 3.23
Timing diagram of an NDSR [23].

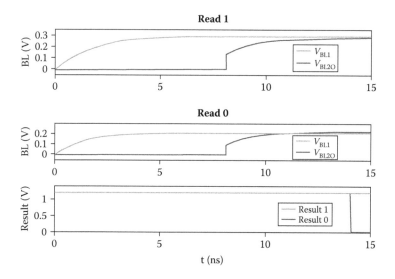

FIGURE 3.24
Functional simulation of an NDSR [23].

3.6.2.4 Voltage-Driven Nondestructive Self-Reference Sensing Scheme

A CSR provides a high sense margin by scarifying the chip reliability, power, and performance. NDSR, however, improves the read latency and power consumption with a smaller sense margin. A voltage-driven nondestructive self-reference sense scheme (VDRS) [44] can provide a trade-off of the

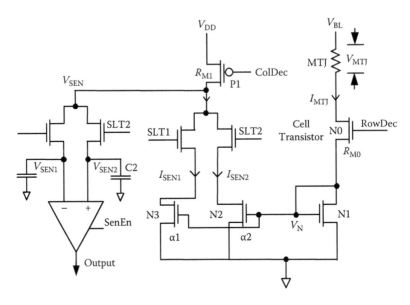

FIGURE 3.25
Voltage-driven nondestructive self-reference sense scheme [44].

nonvolatility, low power consumption, and short latency of an NDSR vs. the high sense margin of a CSR.

The VDRS circuit design is illustrated in Figure 3.25. Similar to NDSR, VDRS relies on the different current dependences of an MgO-based MTJ at two resistance states. However, instead of applying two read currents and comparing the corresponding BL voltages, VDRS applies two different voltages on BL (V_{BL}) and detects the differences between the corresponding currents through the MTJ (I_{MTJ}). The operation of a VDRS is as follows:

First read: An input BL voltage V_{BL1} is applied on an STT-RAM cell and generates the current I_{MTJ1}. A cascode current mirror with a current gain of α amplifies I_{MTJ1} to $I_{SEN1} = \alpha \cdot I_{MTJ1}$. Then I_{SEN1} is translated into a sense voltage $V_{SEN1} = V_{DD} - I_{SEN1} \cdot R_{M1}$. Here R_{M1} is the resistance of the transistor P1 shown in Figure 3.25, which always works at the linear region. V_{SEN1} can be stored in a capacitor C1.

Second read: Apply another input voltage V_{BL2} that is larger than V_{BL1}. Similar to the first read, a sense voltage V_{SEN1} is generated from the current through the MTJ I_{MTJ2} by the other set of cascode current mirrors with a current gain of α. Here $V_{SEN1} = V_{DD} - I_{SEN2} \cdot R_{M1}$. V_{SEN2} can be stored in the other capacitor C2. Here α_2 is carefully chosen to ensure that $V_{SEN1,L} < V_{SEN2,L}$ and $V_{SEN2H} < V_{SEN1,H}$ or:

$$I_{SEN,L1} = \alpha_1 \cdot I_{MTJ,L1} > I_{SEN,L2} = \alpha_2 \cdot I_{MTJ,L2} \qquad (3.12a)$$

$$I_{SEN,H1} = \alpha_1 \cdot I_{MTJ,H1} < I_{SEN,H2} = \alpha_2 \cdot I_{MTJ,H2} \qquad (3.12b)$$

Here, the subscripts H and L stand for the high- and low-resistance states of an MTJ. The subscripts 1 and 2 stand for the first and the second reads.

Sense: By comparing the voltages stored in C1 and C2, the stored value of an STT-RAM bit can be read out as '0' when $V_{SEN1} < V_{SEN2}$ or '1' when $V_{SEN1} > V_{SEN2}$.

Obviously, VDRS is also a nondestructive sensing scheme. Different functional circuitries at two read steps are selected by turning on the select transistors STL1 and STL2, respectively.

3.6.2.5 Comparison of Different Sensing Schemes

Here, we define *sense margin* as the voltage difference between two inputs of a sense amplifier. Accordingly, the distribution of sense margins among the memory bits based on the analytic models is shown in Figure 3.26 with dashed lines. The corresponding simulation results are also shown with solid lines in Figure 3.26 for comparison. Different colors represent different sensing schemes.

The sense margin distribution of CSS is wide and a large portion of the memory bits fall into the negative sense margin region, because a CSS is not fault tolerable to process variation. Bit-to-bit variations of the MTJ and transistor make the yield of the CSS reduce quickly when the required sense margin

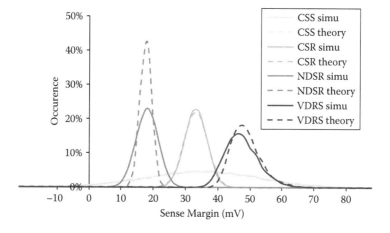

FIGURE 3.26
Sense margin distribution of four sensing schemes [44].

TABLE 3.3

Energy and Latency Comparison
among Four Schemes

Designs	Energy		Latency
	Read '1'	Read '0'	
CSS	0.907 pJ	0.88 pJ	2.5 ns
CSR	22.1 pJ	22 pJ	40 ns
NDSR	1.16 pJ	0.92 pJ	15 ns
VDRS	5.65 pJ	6.43 pJ	15 ns

increases from zero. A CSR has a reasonable yield, that is, 98.1%, even increasing the required sense margin to 20 mV by taking advantage of the large difference between R_H and R_L. An NDSR demonstrates a higher chip yield than a CSS when the required sense margin is small; for example, 8 mV. However, the chip yield sharply drops when the required sense margin is higher than 10 mV. VDRS demonstrates the best performance of process variation tolerance by maintaining a high chip yield until the required sense margin is beyond 45 mV.

A comparison of these four sensing schemes in terms of performance and energy consumption is shown in Table 3.3.

Due to the small read current and short read period, the read energy of a CSS is very small. Compared to a CSS, the read energy of a CSR is much higher because it has two write steps—erase and write back original value. Both NDSR and VDRS have two read steps. The read time period is slightly longer than that in CSS, because capacitor charging requires a certain amount of time; therefore, the read energies of an NDSR must be a little higher than those of a CSS within an acceptable range. VDRS has a relatively higher current flowing through the output path of the current mirrors, which results in high energy consumption.

3.6.3 STT-RAM Architecture Design Technique

The previous sections mainly focused on circuit level design techniques to improve the robustness of an STT-RAM-based memory system. In this section, some architecture-level design techniques will be discussed in order to improve performance, reduce power, and prolong STT-RAM-based memory life span.

3.6.3.1 Read-Before-Write Scheme

One major challenge of STT-RAM is its high write energy, which is much larger than the read energy. According to Zhou et al. [26], the write energy contributes more than 70% of the dynamic energy in an STT-RAM cache, so reducing the write energy of STT-RAM is important to improve its energy efficiency.

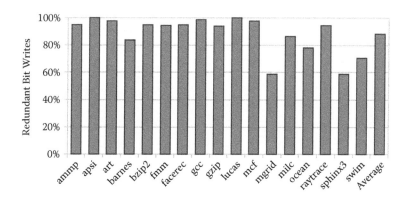

FIGURE 3.27
Redundant bit writes in 16-MB STT-RAM L2 cache [25].

However, in caches of modern computing systems, the write bits and the target bits are same in many write operations. For example, in a 16-MB L2 cache, on average about 88% of bit writes are redundant (see Figure 3.27), which translates to a significant amount of removable bit writes and a great potential for energy savings and performance improvement in an STT-RAM cache [25].

Zhou et al. [26] proposed a read-before-write scheme that can remove the redundant writes to memory. The design diagram is illustrated in Figure 3.28. In this scheme, each column shares one set of decision circuitry that compares the value of the target bit and the write bit to decide whether it is necessary to execute the write. This method requires that every write is preceded by a read operation.

3.6.3.2 Early Write Termination

The write operation in STT-RAM cells has two unique features: (1) the STT-RAM cell still holds a valid old value at the early stage of a write operation [25], and (2) a write operation in STT-RAM is much longer than a read operation. A typical write pulse of an STT-RAM cell is 10 ns, below which the switching current increases rapidly. However, reading from a cell can be completed in less than 1 ns even in a large L2 cache [28].

An early write termination (EWT) method was proposed based on these two unique features [25]. The basic idea of EWT is to sample the resistance of the MTJ (old value) at an early stage of a write operation and turn off the write current if the old value is the same as the new value. The design scheme is shown in Figure 3.29. During the operation of EWT, a write current is applied on a BL to generate a write voltage, which will be compared with the reference to read out the stored value in the cell. If the readout value is same as the one to be written in the cell, the write operation will be terminated by a control signal write cut-off (WCUT). Otherwise, the new value will be written into the cell.

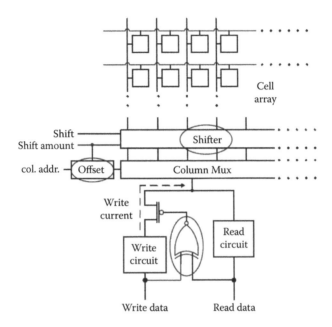

FIGURE 3.28
Implementation of redundant bit-write removal and row shifting [26].

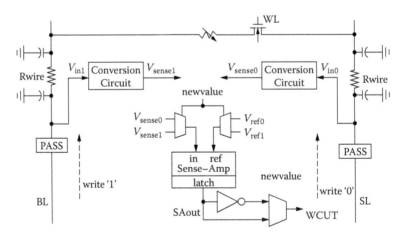

FIGURE 3.29
Early write termination scheme [25].

Compared to the write-after-read scheme, EWT combines a read into the early stage of write operation and, hence, improves the CPU performance. However, an extra sense amplifier might be required because the signal in the read operation in an EWT scheme is generated by a write current, which is much larger than a read current.

3.6.3.3 Data Inverting

Data inverting is another way to reduce the writes to STT-RAM cells [29]. Before writing a new data value into a cache block, the stored data in the target cache block will be read out. Then the Hamming distance (HD) between the two values can be computed. If the calculated HD is larger than half of the cache block size, we invert the new data value ($0 \rightarrow 1$ and $1 \rightarrow 0$) and then store it.

In this method, an extra bit is needed for each cache block to record the inversion status. For instance, the inversion status bit is 1 if the stored value has been inverted. Consequently, during read operations, we need to refer the inversion status bit to recover the original data values. However, if the inversion status bit is 0, which means that the data stored in the cache block have not been inverted, the data can be read out directly. Data inverting can be implemented with a write-after-read scheme to maximize the elimination of redundant write.

3.6.3.4 Wear Leveling

Although the endurance of STT-RAM is expected to be up to 10^{15} programming cycles, the best endurance test result for STT-RAM devices so far is less than 4×10^{12} cycles [30]. The techniques discussed in previous sections, such as read-before-write, early write termination, etc., can prolong the memory life time by eliminating unnecessary writes. However, the write pattern to a memory system is heavily unbalanced. Figure 3.30 demonstrates such an example, which is the write distribution inside an L2 cache when executing

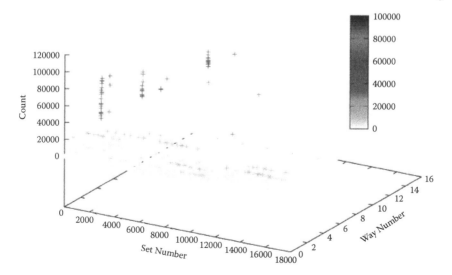

FIGURE 3.30
Write distribution inside an L2 cache.

a SPEC 2006 mcf benchmark. Note that the lifetime of a memory system is determined not by the average write number of all of the memory cells but by several hot cells that are frequently written and wear out ahead of all other cells.

Wear leveling is a method that can help to prolong the memory system lifetime by evening out the write distribution of a memory system; for example, cache or main memory. A wear leveling scheme can be implemented at different granularities and has been studied extensively for flash-based memory or PCM-based memory [26, 31, 32, 41]. These wear leveling schemes can also be implemented in STT-RAM-based storage. The most common wear leveling scheme uses tables to record the access counts on each access unit; for example, a block. Periodically, the most accessed blocks are logically remapped to the least accessed blocks. A separate table is used to record the logical to physical mapping information of these blocks. Usually a wear leveling scheme suffers from significant storage overhead and increasing latency.

In a memory system—for example, a cache in a computing system—we can periodically remap the requested addresses to shift the cache accesses to new physical locations and reduce the write access of some hot locations. For example, we can remap the index bits to achieve row shifting. The index bits can be remapped through certain logic functions—for example, exclusive OR (XOR) to a group of remapping control bits—as shown in Figure 3.28. The remapping control bits can be stored in a nonvolatile register and changed periodically. Another example is simply shifting all of the rows of an array with a given interval. In such a case, a counter needs to be used to record the shift offset. Considering that each shift involves a large amount of data rewrite, the shift interval should be carefully selected. Details of page-level swapping and bit-level shifting are discussed in Zhou et al. [26].

3.6.3.5 Dual Write-Speed Method

The aforementioned schemes can improve the robustness, energy performance, and reliability of an STT-RAM-based memory system by degrading performance. Xu et al. [33] proposed a dual write-speed technique to reduce performance degradation when using small NMOS transistors in STT-RAM cache memory cells.

In STT-RAM design, the MTJ write current follows a distribution for a target MTJ switching time $T_{\text{switch}}^{(t)}$. The size of the NMOS selection transistor is selected to provide the worst-case MTJ write current. Because a larger MTJ write current enables a shorter MTJ switching time, under such a conservative design scenario, most STT-RAM cells can actually enable a switching time (much) shorter than $T_{\text{switch}}^{(t)}$, particularly as the technology scales down. Intuitively, this provides great potential to reduce the *average* STT-RAM cache write latency. Let $T_{\text{switch}}^{(t)}$ denote the target worst-case MTJ switching time and give one design parameter $T_{\text{switch}}^{(s)} < T_{\text{switch}}^{(t)}$, then all of the cache blocks can be partitioned into two categories:

- *Fast Cache Block*: If the shortest switching time allowed by one cache block is not larger than $T_{\text{switch}}^{(s)}$, this cache block is called a *fast cache block* and we use $T_{\text{switch}}^{(s)}$ as the MTJ switching time for this cache block.
- *Slow Cache Block*: If the shortest switching time allowed by one cache block is larger than $T_{\text{switch}}^{(s)}$, this cache block is called a *slow cache block* and we use $T_{\text{switch}}^{(t)}$ as the MTJ switching time for this cache block.

As a result, STT-RAM cache memory only needs to support two different write modes; that is, a fast write mode and a slow write mode, corresponding to fast cache blocks and slow cache blocks. Accordingly, we need to embed a write mode flag memory to store the one-bit write mode configuration information associated with each cache block. Hence, as illustrated in Figure 3.31, to carry out each cache write operation, we first fetch the corresponding write mode flag bit, based on which we execute either a fast cache write with the MTJ switching time of $T_{\text{switch}}^{(s)}$ or a slow cache write with the MTJ switching time of $T_{\text{switch}}^{(t)}$.

To implement this dual-write design strategy, the key design issue is how to choose the appropriate threshold $T_{\text{switch}}^{(s)}$ that can achieve the minimal average cache write latency. The trade-offs among the average MTJ switching time, the probability of the fast cache block, and the threshold $T_{\text{switch}}^{(s)}$ are shown in Figure 3.32.

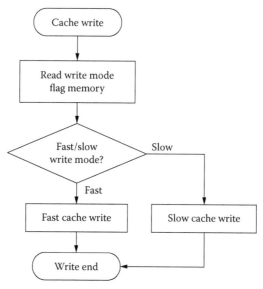

FIGURE 3.31
Cache write operation flowchart of the proposed dual write-speed method [33].

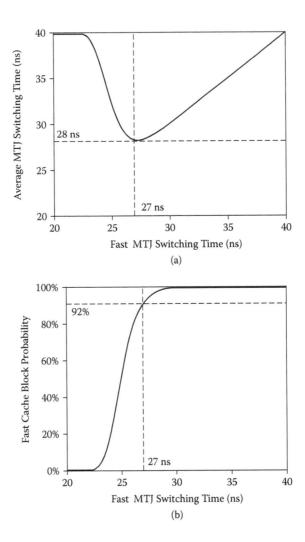

FIGURE 3.32
(a) Average MTJ switching time versus the fast MTJ write switching time $T_{switch}^{(s)}$ and (b) the probability of fast cache blocks versus $T_{switch}^{(s)}$ [33].

3.7 Multilevel Cell

Recently, a two-bit multilevel cell (MLC) MTJ device was demonstrated and reported in Lou et al. [21]. Two-digit information—00, 01, 10, and 11—is represented by four MTJ resistance states. The transitions among different MTJ resistance states are realized by passing the spin-polarized currents with different amplitudes and/or directions. There are two types of MLC MTJs: parallel MLC MTJs and serial MLC MTJs.

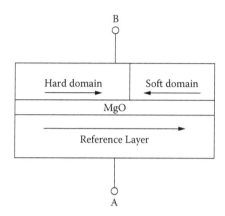

FIGURE 3.33
Parallel MLC MTJ [48].

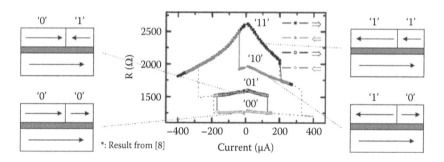

FIGURE 3.34
Four resistance states of MTJ and *R-I* swap curve [48].

Figure 3.33 shows the structure of a parallel MLC MTJ [21]. The free layer of the MLC MTJ has two magnetic domains, of which the magnetic direction (MD) can be changed separately. The MD of one domain (soft domain) can be switched by a small current, whereas that of the other domain (hard domain) can be switched by only a large current. Four combinations of the MDs of the two domains in the free layer correspond to the four resistance states of the MTJs, as shown in Figure 3.34 [49]. The first and the second digits of the two-bit data refer to the MDs of the hard domain and the soft domain, respectively.

Figure 3.35 shows the structure of a serial MLC MTJ. Similar to a parallel MLC MTJ, four levels of resistance can be obtained by combining each state of two MTJs. This stacked structure of two MTJs can be fabricated by a self-aligned process without additional lithography [22]. An illustration of the fabrication process is shown in Figure 3.36.

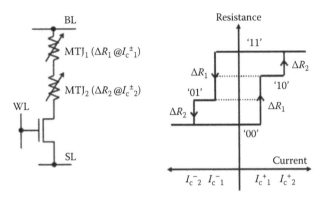

FIGURE 3.35
Serial MLC MTJ and its *I-R* curve [22].

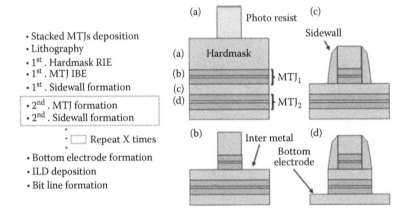

FIGURE 3.36
Fabrication process of serial MLC MTJ [22].

3.7.1 Write Scheme for Multilevel Cells

In the write operation of MLC STT-RAM cells, a write current with different amplitudes and directions is applied to the MTJ by adjusting the biases on the WL and BL. Table 3.4 shows the switching currents required by the transitions between the different resistance states of a parallel MLC MTJ with a 100 nm × 200 nm elliptical shape. A positive current denotes the current direction from the free layer to reference layer. Although the MD of the soft domain can be switched alone, switching of the MD of the hard domain is always associated with the switching of the MD of the soft domain. In general, switching the MD of the soft domain requires a smaller current than that required to switch the MDs of both hard and soft domains.

Transitions between some pairs of resistance states cannot complete directly (shown as "X" in Table 3.4). For example, the transition 11 → 01 has to

TABLE 3.4

Switching Currents of MLC STT-RAM Cell (uA)

From	To			
	00	01	10	11
00	0	−189	X	−280
01	130	0	X	−280
10	328	X	0	−45
11	328	X	197	0

go through two steps: the first step is from 11 to 00 by applying a switching current of 328 µA; the second step is from 00 → 01 by applying a switching current of −189 µA. '0' denotes the switching currents required by the transitions between the same states.

Without loss of generality, the transitions of the MTJ resistance states can be summarized as the following four types [48]:

1. Zero transition (ZT): the MTJ stays in the same state.
2. Soft transition (ST): Only the MD of the soft domain is switched in the transition; that is, 00 → 01 and 10 → 11.
3. Hard transition (HT): The MDs of both soft and hard domains are switched in the transition; that is, 00 → 11, 01 → 11, 10 → 00, and 11 → 00.
4. Two-step transition (TT): Transition completes with two steps, including one HT followed by one ST; that is, 00 → 10, 01 → 10, 10 → 01, and 11 → 01.

There are three write schemes of MLC STT-RAM cells—simple, complex, and hybrid. The state transition graphs of these three write schemes are summarized in Figure 3.37.

3.7.1.1 Simple Write Scheme of MLC STT-RAM Cells

Depending on the bit that is being written, a simple write scheme of an MLC STT-RAM can be proposed as follows:

- Writing 00 and 11: the MLC STT-RAM bit is directly programmed to the state by one HT.
- Writing 01 and 10: a TT is executed; the MLC STT-RAM bit is first programmed to 00 or 11 by one HT and then programmed to 01 or 10 by one ST.

We refer to this write scheme as a *simple write scheme*.

3.7.1.2 Complex Write Scheme of MLC STT-RAM Cells

In fact, many unnecessary transitions are introduced in simple write schemes. For example, for the transition 10 → 11, a TT is executed in the simple write scheme even if it can be completed only by an ST. To solve this issue, we proposed a complex write scheme as follows: A read operation is conducted first. Based on the values of the new data being written and the original data stored in the MLC STT-RAM bit, a ZT, ST, HT, or TT will be executed exactly according to the transition procedure shown in Table 3.4.

3.7.1.3 Hybrid Write Scheme of MLC STT-RAM Cells

The power dissipation of an HT is significantly higher than that of an ST. We noticed that the transitions between the two resistance states with the same first digit can be completed by only one ST; that is, 00 → 01 and 10 → 11. Also, the first digit of an MLC STT-RAM bit value can be read by only one comparison between V_{BL} and V_{REF2}. Based on these two observations, we propose a hybrid write scheme to reduce the power dissipation by minimizing the write operations that require an HT. The write operation types of hybrid write schemes can be summarized as follows:

- Soft hybrid write (SH): If the first digit of the new data and the original data are the same, only one ST is executed to complete the transition; that is, 00 → 01, 10 → 11, 00 → 00, and 11 → 11.

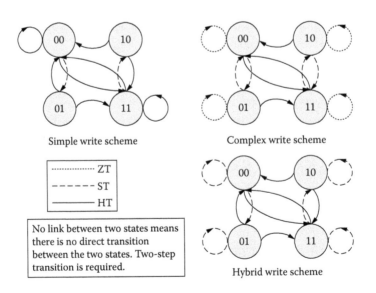

FIGURE 3.37
State transition graphs of write schemes [48].

- Hard hybrid write (HH): If the first digit of the new data and the original data are different, and the new data is either 00 or 11, only one HT is needed to complete the transition; that is, $00 \to 11, 01 \to 11, 10 \to 00$, and $11 \to 00$.

- Two-step hybrid write (TH): If the first digit of the new data and the original data are different, and the new data is either 01 or 10, transition completes with a TT; that is, $00 \to 10, 01 \to 10, 10 \to 01$, and $11 \to 01$.

For random inputs, the probabilities of SH, HH, and TH are one half, one quarter, and one quarter, respectively.

3.7.2 Sensing Scheme for MLC

To read the data from an MLC STT-RAM cell, a multireference sense amplifier for the two-step read operation is required. A dichotomic read scheme

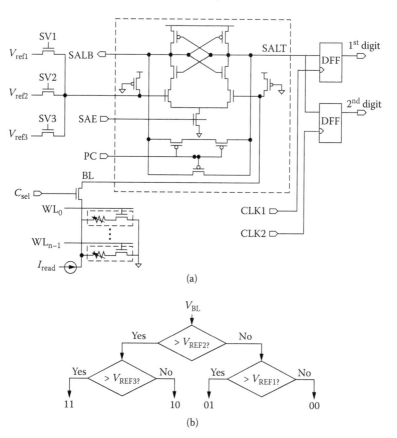

(a)

(b)

FIGURE 3.38
Dichotomic read scheme of MLC STT-RAM: (a) schematic and (b) decision flow [48].

of an MLC STT-RAM is shown in Figure 3.38. This design scheme requires three references, each of which is between two neighboring resistance sates. The most significant bit (MSB) and least significant bit (LSB) are read by using two read cycles. In the first step, the sense amplifier uses V_{REF2} as a reference to identify the MSB. In the second step, to read the LSB, the reference resistance is switched to V_{REF1} or V_{REF3}, depending on the MSB value. In this read circuitry, only one sense amplifier is required.

References

[1] Berger, L. (1996). Emission of spin waves by a magnetic multilayer traversed by a current. *Physical Review B*, 54: 9353–9358.

[2] Slonczewski, J. C. (1996). Current-driven excitation of magnetic multilayers. *Journal of Magnetism and Magnetic Materials*, 159: L1–L7.

[3] Stiles, M. D. and Zangwill, A. (2002). Anatomy of spin-transfer torque. *Physical Review B*, 66: 014407.

[4] Zhang, S., Levy, P. M., and Fert, A. (2002). Mechanisms of spin-polarized current-driven magnetization switching. *Physical Review Letters*, 88: 236601.

[5] Katine, J. A., Albert, F. J., Buhrman, R. A., Myers, E. B., and Ralph, D. C. (2000). Current-driven magnetization reversal and spin-wave excitations in Co/Cu/Co pillars. *Physical Review Letters*, 84: 3149–3152.

[6] Jiang, Y., Abe, S., Ochiai, T., Nozaki, T., Hirohata, A., Tezuka, N., and Inomata, K. (2004). Effective reduction of critical current for current-induced magnetization switching by a ru layer insertion in an exchange-biased spin valve. *Physical Review Letters*, 92: 167204.

[7] Liu, Y., Zhang, Z., Freitas, P. P., and Martins, J. L. (2003). Current-induced magnetization switching in magnetic tunnel junctions. *Applied Physics Letters*, 82: 2871–2873.

[8] Deac, A., Redon, O., Sousa, R. C., Dieny, B., Nozières, J. P., Zhang, Z., Liu, Y., and Freitas, P. P. (2004). Current driven resistance changes in low resistance x area magnetic tunnel junctions with ultra-thin Al-Ox barriers. *Journal of Applied Physics*, 92: 6792–6794.

[9] Chen, Y., Wang, X., Li, H., Xi, H., Zhu, W., and Yan, Y. Design margin exploration of spin-transfer torque RAM (STT-RAM) in scaled technologies. *IEEE Transactions on Very Large Scale Integration Systems*, 18: 1724–1734.

[10] Huai, Y., Albert, F., Nguyen, P., Pakala, M., and Valet, T. (2004). Observation of spin-transfer switching in deep submicron-sized and low-resistance magnetic tunnel junctions. *Applied Physics Letters*, 84: 3118–3120.

[11] Fuchs, G. D., Emley, N. C., Krivorotov, I. N., Braganca, P. M., Ryan, E. M., Kiselev, S. I., Sankey, J. C., Katine, J. A., Ralph, D. C., Buhrman, R. A., and Katine, J. A. (2004). Spin-transfer effects in nanoscale magnetic tunnel junctions. *Applied Physics Letters*, 85: 1205–1207.

[12] Chen, Y., Wang, X., Li, H., Liu, H., and Dimitrov, D. V. (2008). Design margin exploration of spin-torque transfer RAM (SPRAM). Paper presented at the International Symposium on Quality Electronic Design, Santa Clara, CA.

[13] Kawahara, T., Takemura, R., Miura, K., Hayakawa, J., Ikeda, S., Lee, Y., Sasaki, R., Goto, Y., Ito, K., Meguro, I., Matsukura, F., Takahashi, H., Matsuoka, H., and Ohno, H. (2008). 2 Mb SPRAM (spin-transfer torque RAM) with bit-by-bit bi-directional current write and parallelizing-direction current read. *IEEE Journal of Solid-State Circuits*, 43: 109–120.

[14] Chen, Y., Li, H., Wang, X., Zhu, W., Xu, W., and Zhang, T. (2010). A nondestructive self-reference scheme for spin-transfer torque random access memory (STT)-RAM. design. Paper presented at the Automation & Test in Europe Conference and Exhibition, Dresden, Germany.

[15] Wang, X., Zhu, W., Siegert, M., and Dimitrov, D. (2009). Spin torque induced magnetization switching variations. *IEEE Transactions on Magnetics*, 45: 2038–2041.

[16] Hosomi, M., Yamagishi, H., Yamamoto, T., Bessho, K., Higo, Y., Yamane, K., Yamada, H., Shoji, M., Hachino, H., Fukumoto, C., Nagao, H., and Kano, H. (2005). A novel nonvolatile memory with spin torque transfer magnetization switching: Spin-RAM. Paper presented at the IEEE International Electron Devices Meeting (IEDM), Washington, DC.

[17] Beech, R., Anderson, J., Pohm, A., and Daughton, J. (2000). Curie point written magnetoresistive memory. *Journal of Applied Physics*, 87: 6403.

[18] Simmons, J. G. (1963). Generalized formula for the electric tunnel effect between similar electrodes separated by a thin insulating film. *Journal of Applied Physics*, 34: 1793–1803.

[19] Sun, J. Z. (2000). Spin-current interaction with a monodomain magnetic body: A model study. *Physical Review B*, 62: 570.

[20] Serpico, C., d'Aquino, M., Bertotti, G., and Mayergoyz, I. D. (2005). Analytical approach to current-driven self-oscillations in Landau-Lifshitz-Gilbert dynamics. *Journal of Magnetism and Magnetic Materials*, 290: 502–505.

[21] Wang, X., Zheng, Y., Xi, H., and Dimitrov, D. (2008). Thermal fluctuation effects on spin torque induced switching: Mean and variations. *Journal of Applied Physics*, 103: 034507.

[22] Koch, R. H., Katine, J. A., and Sun, J. Z. (2004). Time-resolved reversal of spin-transfer switching in a nanomagnet. *Physical Review Letters*, 92: 088302.

[23] Higo, Y., Yamane, K., Ohba, K., Narisawa, H., Bessho, K., Hosomi, M., and Kano, H. (2005). Thermal activation effect on spin transfer switching in magnetic tunnel junctions. *Applied Physics Letters*, 87: 082502.

[24] Li, J., Liu, H., Salahuddin, S., and Roy, K. (2008). Variation-tolerant spin-torque transfer (STT) MRAM array for yield enhancement. Paper presented at the IEEE Custom Integrated Circuits Conference (CICC) San José, CA.

[25] Li, J., Augustine, C., Salahuddin, S., and Roy, K. (2008). Modeling of failure probability and statistical design of spin-torque transfer magnetic random access memory (STT MRAM) array for yield enhancement. *Proceedings DAC*, 278–283.

[26] Li, J., Ndai, P., Goel, A., Salahuddin, S., and Roy, K. (2010). Design paradigm for robust spin-torque transfer magnetic RAM (STT MRAM) form circuit/architecture perspective. *IEEE Transactions on Very Large Scale Integration Systems*, 18: 1710–1723.

[27] Prejbeanu, I. L., Kerekes, M., Sousa, R. C., Sibuet, H., Redon, O., Dieny, B., and Nozières, J. P. (2007). Thermally assisted MRAM. *Journal of Physics: Condensed Matter*, 19: 165218.

[28] Prejbeanu, I. L., Kula, W., Ounadjela, K., Sousa, R.C., Redon, O., Dieny, B., and Nozieres, J.P. (2004). Thermally assisted switching in exchange-biased storage layer magnetic tunnel junctions. *IEEE Transactions on Magnetics*, 40: 2625–2627.

[29] Xi, H., White, R. M., Gao, Z., and Mao, S. (2002) Antiferromagnetic thickness dependence of blocking temperature in exchange coupled poly-crystalline ferromagnet/antiferromagnet bilayers. *Journal of Applied Physics*, 92: 4828–4831.
[30] Li, H., Xi, H., Chen, Y., Wang, X., and Zhang, T. (2009). Thermal-assisted spin transfer torque memory (STT-RAM) cell design exploration. Paper presented at the IEEE Computer Society Annual Symposium on VLSI (ISVLSI), Tampa, FL.
[31] Wang, J. and Freitas, P. P. (2004). Low-current blocking temperature writing of double-barrier MRAM cells. *IEEE Transactions on Magnetics*, 40: 2622–2624.
[32] Nogues, J. and Schuller, I. K. (1999). Exchange bias. *Journal of Magnetism and Magnetic Materials*, 192: 203–232.
[33] Daughton, J. M. and Pohm, A. V. (2003). Design of Curie point written random access memory cells. *Journal of Applied Physics*, 93: 7304–7306.
[34] Sun, J. Z., Allensopach, R., Parkin, S., Slonczewski, J. C., and Terris, B. D. (2005). *Spin-current switched magnetic memory element suitable for circuit integration and method of fabricating the memory element.* U.S. Patent Application 20050104101.
[35] Kiselev, S. I., Sankey, J. C., Krivorotov, I. N., Emley, N. C., Schoelkopf, R. J., Buhrman, R. A., and Ralph, D. C. (2003). Microwave oscillations of a nanomagnet driven by a spin-polarized current. *Nature*, 425: 380–383.
[36] Foster, H., Bertram, N. H., Wang, X., Dittrich, R., and Schrefl, T. (2003). Energy barrier and effective thermal volume in columnar grains. *Journal of Applied Physics*, 267(1): 69–79.
[37] Capodieci, L. (2006). From optical proximity correction to lithography-driven physical design (1996–2006): 10 years of resolution enhancement technology and the roadmap enablers for the next decade. *Proceedings SPIE*, 6154: 615401.
[38] Sun, Z., Li, H., Chen, Y., and Wang, X. (2010). Variation tolerant sensing scheme of spin-transfer torque memory for yield improvement. Paper presented at the IEEE International Conference on Computer-Aided Design, San José, CA.
[39] Zhou, P., Zhao, B., Yang, J., and Zhang, Y. (2009). A durable and energy efficient main memory using phase change memory technology. Paper presented at the 36th International Symposium on Computer Architecture, Austin, TX.
[40] Zhou, P., Zhao, B., Yang, J., and Zhang, Y. (2009). Energy reduction for STT-RAM using early write termination. Paper presented at the IEEE/ACM International Conference on Computer-Aided Design, San José, CA.
[41] Dong, X., Wu, X., Sun, G., Xie, Y., Li, H., and Chen, Y. (2008). Circuit and microarchitecture evaluation of 3D stacking magnetic RAM (MRAM) as a universal memory replacement. Paper presented at the Design Automation Conference, Anaheim, CA
[42] Joo, Y., Niu, D., Dong, X., Sun, G., Chang, N., and Xie, Y. (2010). Energy- and endurance-aware design of phase change memory caches. Paper presented at the Design, Automation & Test in Europe Conference & Exhibition, Dresden, Germany.
[43] Croes, K. and Tokei, Z. (2010). E- and $\sqrt{E}$- model too conservative to describe low field time dependent dielectric breakdown. Paper presented at the 2010 International Reliability Physics Symposium, Anaheim, CA.
[44] Gal, E. and Toledo, S. (2005). Algorithms and data structures for flash memories. *ACM Computing Surveys*, 37(2): 138–163.
[45] Kgil, T., Roberts, D., and Mudge, T. (2008). Improving nand flash based disk caches. Paper presented at the 35th Annual International Symposium on Computer Architecture, Beijing, China.

[46] Quresh, M. K., Karidis, J., Franceschini, M., Srinivasan, V., Lastras, L., and Abali, B. (2009). Enhancing life time and security of PCM-based main memory with start-gap wear leveling. Paper presented at the International Symposium on Microarchitecture, New York.

[47] Xu, W., Sun, H., Wang, X., Chen, Y., and Zhang, T. (2009). Design of last-level on-chip cache using spin-torque transfer RAM (STT RAM). *IEEE Transactions on Very Large Scale Integration Systems*, 19: 1–11.

[48] Lou, X., Gao, Z., Dimitrov, D. V., and Tang, M. X. (2008). Demonstration of multilevel cell spin transfer switching in MgO magnetic tunnel junctions. *Applied Physics Letters*, 93: 242502.

[49] Chen, Y., Li, H., Sun, Z., Wang, X., Zhu, W., Sun, G., and Xie, Y. (2010). Access scheme of multi-level cell spin-transfer torque random access memory and its optimization. Paper presented at the 53rd IEEE International Midwest Symposium on Circuits and Systems (MWSCAS), Seattle, WA.

[50] Ishigaki, T., Kawahara, T., Takemura, R., Ono, K., Ito, K., Matsuoka, H., and Ohno, H. A multi-level-cell spin-transfer torque memory with series-stacked magneto-tunnel junctions. Paper presented at the 2010 Symposium on VLSI Technology (VLSIT), Honolulu, HI.

4

Resistive Random Access Memory

In general, resistive random access memory (R-RAM) denotes all of the random access memories that rely on the resistance differences to store data. Usually, a R-RAM storage device has a sandwich metal–insulator–metal (MIM) structure. By changing the insulator layer to a high-resistance state (HRS) or a low-resistance state (LRS), a R-RAM device can represent logic '0' or '1,' respectively. Various R-RAM materials based on different physical mechanisms have been extensively studied. For its high density, low power consumption, simple fabrication process, and good scalability, R-RAM has become a good candidate to substitute for traditional data storage technologies (e.g., hard disk drives and flash memory) [5, 7].

Because R-RAM relies on the resistance difference of the storage device to differentiate stored logic value, sometimes we also call the storage device a *resistance switching cell*. Usually it is a MIM structure that has a tunable insulator sandwiched in between two metal electrodes. The switching characteristic has on and off states corresponding to logic '0' and '1,' respectively. Increasing the number of resistance levels can achieve more logic states in one R-RAM cell. Such a memory cell is called a *multilevel cell* (MLC).

4.1 Variety of R-RAM

There are many different R-RAM materials with different switching mechanisms. In this section, we will briefly introduce some popular nano-ionic mechanisms classified by materials.

4.1.1 Electrochemical Metallization Devices

Filament is a key word when describing the conduction of R-RAM. A filament is not a traditional conducting material such as a metal or a transportation carrier of a semiconductor. The filaments inside an insulator are the tiny paths that connect the two metal electrodes. The filaments only occupy a small portion of the isolator and the rest remain nonconductive.

In a nano-ionic R-RAM, the filament can appear or vanish by controlling the bias condition on the device electrodes.

In a MIM structure, the metal of one electrode is inert, such as Au, Pt, etc. The active metals, such as Cu, Al, Ni, etc., are used for the other electrode, which takes part in the process of forming filaments: the biasing on the electrodes forces the active metal ionizations to migrate into the insulator and combine with the dissolving ion within the insulator layer. As a result, a stable compound with conducting characteristics forms. To break the weak chemical bond of the insulator and then form a stable energy state compound in the forming procedure, a continuous voltage or current needs to be supplied. After there are enough conducting compounds in the insulator, filaments form from the active electrode to the inert electrode. The electrons can freely move across filament and conduct current between the two electrodes. Thus, the device changes to a low-resistance state. Similarly, for some insulators made of electrolytes, the cation reacts with the electrolytes to form a filament.

An electrochemical metallization device requires three steps for a set procedure:

1. With the applied voltage, the device incurs anodic M dissolution and then generates M^{Z+}, which is metal cations in the electrolyte, and a corresponding amount of (Z) electrons.

$$M \rightarrow M^{Z+} + Ze^-$$

2. M^{Z+} migrates across solid electrolyte under an electrical field.
3. Reduction and electrocrystallization of M occur at the inert electrode and form a filament, such as

$$M^{Z+} + Ze^- \rightarrow M.$$

Afterwards, the filament grows from an inert electrode to an active electrode and creates a path for electronics transportation. The R-RAM cell is then in an LRS. The current density of the device is determined by the number of filaments within the insulator. In the reset procedure, an opposite bias condition is applied across the R-RAM to dissolve the conducting filament back to its original status.

In some applications, the M^+ is prefabricated into an electrolyte layer. This kind of device does not go through a dissolution step during a set procedure. As a result, the energy consumption of the forming filament can be reduced and it can achieve a faster set operation. The set time is also affected by the filament-forming procedure and the migration speed of M^+ in the electrolyte. In general, a high electrical field can speed up a set procedure.

A device with an electrochemical metallization (ECM) mechanism can be used as a memory cell or switching device in R-RAM application. Figure 4.1 is an example of an ECM R-RAM cell with GeSe as the electrolyte, Ag as the active metal, and Pt as the inert metal [6, 11]. The Ag and GeSe electrolyte form the filament for carrier transportation.

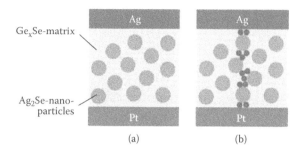

FIGURE 4.1
Sketch of an Ag/Ag-Ge-Se/Pt cell with nanodispersed Ag2Se particles in a Ge-Se matrix. (a) OFF state; (b) ON state. [6]

4.1.2 Valence Change System

A valence change system can be classified into two types: (1) the charge trapping mechanism, which modifies the electrostatic barrier in the MIM structure, and (2) the insulator metal transition (IMT), in which the electronic is injected into the insulator layer to conduct the current.

The electrochemical metallization introduced in Section 4.2.1 utilizes a chemical mechanism to set or reset the R-RAM device. Conversely, the valence charge system is based only on the device's physical mechanism and it does not form a filament in the insulator. When applying a bias across the device, pseudo-states are added between the valence band and conduction band. The electrons can jump through these states from the valence band to the conduction band. The charge trapping system applies a high electrical field to force the charges to migrate into the insulator through Fowler–Nordheim tunneling. These charges could be trapped by defects in the insulator. Thus, the barrier at the interface of metal and insulator is modified to result in Schottky barrier lowering and Schottky barrier tunneling. Indeed, popular storage device flash memories utilize a similar mechanism. The barrier height at the junction of the metal-to-insulator layer can be changed by supplying voltage across the device. As a result, the threshold voltage can be modified to represent different logic values.

The IMT is another valence change system. Under applied electrical field, the electronic charge is injected into the insulator. The charge injection in a device with a superlattice structure and a disordered lattice and mixed valence band behaves like an ion-doping process. A typical device with an IMT mechanism is $Pr_{0.7}Ca_{0.3}MnO_3$ (PCMO).

As shown in Figure 4.2, PCMO is a superlattice structure with different band gaps. The structure of a PCMO device can be separated into three parts: conduction metal oxide (CMO) in the superlattice structure; TO, the tunnel oxide at the interface; and Pt, the inert electrode. By tuning the barrier height of TO through an electrical field and oxygen migration, the probability of the tunneling carrier to conduct current can be increased or reduced and, hence, the PCMO resistance changes.

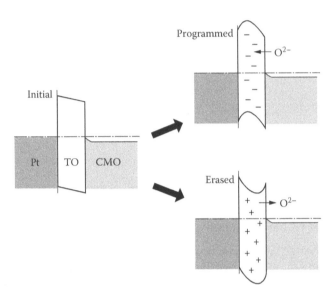

FIGURE 4.2
Operating scheme of PCMO. The program operation and erase operation modify the barrier height of TO [6].

PCMO is a p-type valence change device and its biasing condition in set/ reset operations is opposite of that of ECM devices. PCMO is an active LRS material and therefore electrons can tunnel through Schottky barrier. A low work function metal—that is, Ti—is used as the electrode in PCMO devices to form a Schottky barrier.

Figure 4.2 shows the barrier height variation of PCMO under forward and reverse biasing. CMO is made of superlattice structure that has a mixed density of state. Part of the CMO acts as conductor and the remainder combines with Ti and forms tunnel oxide. The defect-induced carrier migration results in the LRS status. In a program operation, oxygen ions driven by a large electrical field are injected into the oxide layer. As a result, the defects in the TO reduces. When a reverse bias is applied, the migrating oxygen ion is expelled out of the oxide during erase operations. The extra state in the oxide layer provides electronics to jump from the valence band to the conduction band. The operating voltage of PCMO is normally higher than that of ECM. Unlike ECM devices, which form filaments for carrier transportation, the insulator metal transition is a surface change-inducing carrier transportation under high electrical field.

Another example of a valence change system material is n-type $SrTiO_3$, which has an active HRS. Its operating scheme is similar to that of ECM devices, but the switching mechanism is barrier modification. Whether a device is p-type or n-type depends on the type of work function of the inert metal.

4.2 R-RAM Operations

Set and reset are two basic R-RAM operations. Based on the various switching mechanisms introduced in Section 4.1, the biasing conditions supplied to the different R-RAM materials, and hence the corresponding operation schemes, could be different.

From the programming point of view, all R-RAM materials fall into only two operation types—unipolar switching and bipolar switching. Figures 4.3(a) and 4.3(b) show the *I-V* characteristics of unipolar and bipolar R-RAM designs, respectively.

Unipolar operation executes programming/erasing by using short/long pulses or by using high/low voltage with the same voltage polarity. The reverse biasing condition is prohibited. Some unipolar R-RAM designs connect the storage element with a diode to eliminate the bias voltage or current in the unneeded direction. To switch resistance status, the voltage of a set operation (programming) is always higher than that of a reset operation (erasing), and the reset current is always higher than the set current. Ideally, the energy consumed in a set operation and a reset operation should be equivalent. Many materials have demonstrated unipolar switching characteristics, such as phase change memory (PCM). The major design difficulty in a unipolar R-RAM is accurately controlling the pulse width for set and reset operations.

The operations of a bipolar switching R-RAM design are simple. Short pulses with opposite voltage polarities are supplied for set and reset operations as shown in Figure 4.3(b). Note that the absolute value of the set voltage does not have to the same as that of the reset voltage. This is true for most of R-RAM materials.

Usually, we use set to describe the state transition from HRS to LRS. Reset is a reverse procedure to a set operation. However, the forward biasing is not always associated with a set operation and a reset operation does not

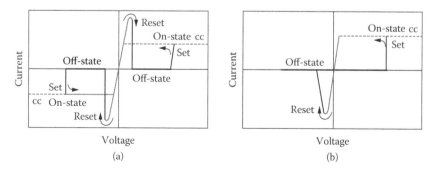

FIGURE 4.3
(a) Unipolar operation of R-RAM and (b) bipolar operation of R-RAM [6].

always require a reverse bias for all devices. Based on the different mechanism of device conduction, we need to make sure that the correct voltage/current amplitude and polarity can be provided to the storage device.

The threshold voltage is usually defined for R-RAM material that represents the state transition point. Once the supplying voltage surpasses the threshold voltage, the resistance could change abruptly. This is a continuous process, but the transition is usually completed in a short period of time; that is, less than a nanosecond. If there is a switching device connected to the storage element in a R-RAM cell design, a skewed threshold voltage should be considered.

4.3 Single Cell R-RAM Design

Let us start with a single memory cell to explain the design considerations in R-RAM. Due to its high R_{off}/R_{on} ratio, R-RAM can be constructed without any switching device in a memory cell in order to achieve high array density. However, in some applications, a switching device is still preferred for better memory access control. Different R-RAM cell structures will be introduced in this section.

4.3.1 Crossbar Structure Design without Switching Device

A crossbar array is widely used in R-RAM design due to its simple structure and high density. A crossbar was firstly initiated and demonstrated in a telecommunication switching system, which contained two sets of wires and switches at cross points. Signal routing was controlled by properly selecting switches. In nanometer-scale high-density memory design, a similar structure is maintained—a storage element is placed at each cross point of two sets of metal wires [17]. Theoretically, we can achieve the smallest memory cell area of $4F^2$ by using crossbar array structure, where F is the minimum feature size [2].

Figure 4.4(a) illustrates the top view of a R-RAM crossbar structure with $4F^2$ cell area, and Figure 4.4(b) shows the corresponding circuit digram. Here, word line (WL) and bit line (BL) represent the horizonal and vertical interconnects, respectively. To achieve the smallest cell area, the minimal geometry metal wire width and space are required, which is mainly determined by the lithography limitation. As we can see, each side of such a memory cell has a size of $2F$, and the area of a single memory cell is $4F^2$.

The crossbar array design without switching devices has advantages including a simple process and high array density. Moreover, 3D stacking by constructing multiple memory layers vertically benefits from such a design with thin device thickness. However, the crossbar array without switching could form sneak paths, which have three or more cells in series, as shown in Figure 4.4(b). In such a design, the voltage across the selected cell must be

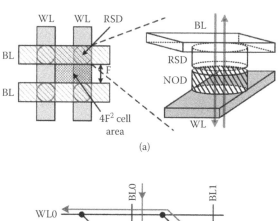

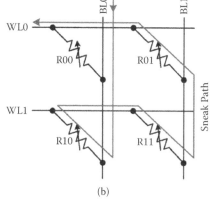

FIGURE 4.4
Crossbar array design without switching device: (a) top view of layout and (b) circuit diagram.

much higher than that across the other cells in the sneak path to guarantee proper functionality. When increasing the crossbar array size, the sensing margin in a read operation can be degraded quickly due to the existence of a sneak path. Accordingly, it requires R-RAM material to have a high resistance in an LRS and a large difference in resistance between the HRS and LRS. For example, the resistance of an LRS should be at least several KW and HRS should be approximately in the order of MW [3, 16].

4.3.2 R-RAM Cell Design with Active Switching Device

Active devices such as a metal-oxide semiconductor field-effect transistor (MOSFET) and bipolar junction transistor (BJT) can be used as switching devices in memory design due to their better controllability. For example, Figure 4.5 demonstrates a R-RAM cell design using a MOSFET as switching device. The gate of a MOSFET transistor is isolated from other terminals and can act as a throttle to turn the current on or off across the R-RAM material. Proper access control with active devices can significantly reduce the leakage current and minimize the impacts of the sneak path in a R-RAM array.

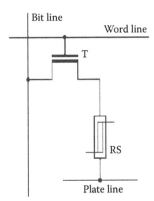

FIGURE 4.5
R-RAM cell design with a MOSFET as the switching device [6].

Therefore, a large memory array can be built by using MOSFETs as switching devices.

The main drawback of this design is the large memory cell area, which is more than $10F^2$. Compared to the $4F^2$ cell area of a crossbar structure, the memory density when using a MOSFET as a switching device is only half or even less. Moreover, as we will introduce in Section 4.5, 3D technology has been proven as a promising technique to improve memory density. But integrating multiple MOSFET layers vertically is constrained by the fabrication process. However, this critical issue potentially could be solved by using the thin-film transistor (TFT) technology, which is currently being investigated by industrial companies [12].

To improve single-layer memory density, vertical BJTs [13] are a potential candidate. Figure 4.6 shows the structure using a vertical BJT with a R-RAM cell upon a substrate. In the illustration, N^+ is beneath the R-RAM cell in the vertical direction and P^+ diffusions are in the horizontal direction as a WL. Note that the process of a vertical BJT is compatible with the modern complementary metal-oxide semiconductor (CMOS) process, which lowers the bar for development and commercialization. Similar to a MOSFET, a vertical BTJ can confine the current through the device and hence achieve good cell selectivity. Moreover, such a R-RAM cell design can obtain a $4F^2$ cell size, which is theoretically the smallest memory cell size on one layer. However, a vertical BJT cannot be stacked up to form a multilayer memory island because the process of P^+ and N^+ diffusion is hard to engineer in the upper layers.

4.3.3 R-RAM Cell Design with Passive Switching Device

In addition to active devices, passive devices, such as diodes and non-ohmic devices (NODs), can be used as switching devices in R-RAM design. Those two-terminal passive devices can be integrated with R-RAM material

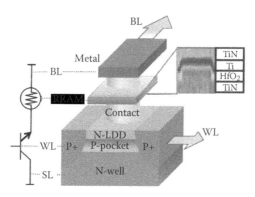

FIGURE 4.6
Vertical BJT structure [13].

during the fabrication process to construct a crossbar structure. For a unipolar R-RAM memory, a diode in series with a data storage cell structure can be used as a switching device (1D1R), as shown in Figure 4.3(a). The switching device in bipolar R-RAM memory can be an NOD, as shown in Figure 4.3(b) [2].

Diodes can be used as the switching device for unipolar R-RAM, as shown in Figure 4.7(a) [14]. A diode turns on only when the driving voltage exceeds its threshold voltage in the forward direction. Otherwise, only a small leakage current can go through a diode. Vertical diodes and heterojunction diodes are promising candidates for R-RAM switching devices because they can help to maintain $4F^2$ of single cell area.

An NOD is a nonlinear device that can be integrated under the device, as shown in Figure 4.7(b) [15]. An NOD is like a varistor (variable resistor) with nonlinearity characteristics, or it can be approximated as a two parallel connected diodes in opposite directions. The device can be turned on no matter what the forward or backward biasing is over its threshold voltage. In such a condition, the target R-RAM cells can be accessed. An NOD can be used as a switching device in bipolar R-RAM design. Because the real device cannot achieve ideal on/off characteristics, the sneak path remains an important design issue to be considered.

Another passive switching device—threshold voltage switching—is one type of R-RAM device that is made of chalcogenide material. A chalcogenide device is an electrochemical metallization device as mentioned in Section 4.1. When utilizing it as a switching device, it can be turned on (e.g., changed to a low resistant state) through a set operation so as to write/read the storage cell. The threshold voltage switching can be naturally integrated with storage elements in the vertical direction to obtain a $4F^2$ cell area. Compared to a diode, the threshold voltage switching device has a lower threshold voltage so that it is easy to turn on. Furthermore, the device can be driven bidirectionally after it enters a low-resistance state. Hence, it fits well with bipolar R-RAM design [11].

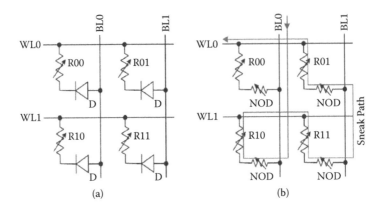

FIGURE 4.7
R-RAM memory array: (a) 1D1R structure and (b) using an NOD as a switching device.

There are certain design considerations when integrating passive switching devices in R-RAM cells. For example, a crossbar array with passive device needs to turn on the target cell while maintaining the unselected cells in off mode. As the array size increases, the conducting current through sneak paths can dominate. As a result, the array size is constrained.

4.4 R-RAM Scalability

In this section, we use a PCMO-based resistive switching device to explain the scalability of R-RAM design.

4.4.1 Basic Memory Structure and Its Fabrication

Figure 4.8 shows the structure of our PCMO-based resistive switching device (RSD): The PCMO film is symmetrically sandwiched between a Pt bottom electrode (BE) and a Pt top electrode (TE). The whole RSD was fabricated on a patterned TiN/Ti substrate. PCMO layers with a thickness of 30 nm were deposited by radio frequency–magnetron sputtering at wafer temperatures between 380 and 410°C. After TE deposition was completed at room temperature, RSDs were patterned by traditional lithography techniques to form isolated cells with various sizes and aspect ratios. Here the aspect ratio of an RSD is the ratio of its longer dimension to its shorter dimension.

Figures 4.9 and 4.10 show the X-ray diffraction (XRD) q–2q scan and the cross-sectional transmission electron microscopy (TEM) image of a sheet film of PCMO, respectively. Figure 4.9 indicates that the PCMO layer is crystalline with a 110 out-of-plane texture. In Figure 4.10, the grains are

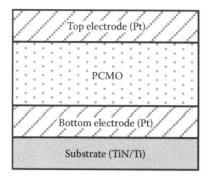

FIGURE 4.8
PCMO-based device structure.

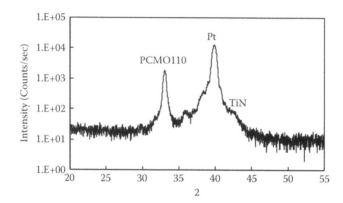

FIGURE 4.9
X-ray diffraction q–2q scan of memory stack with 30-nm PCMO layer. PCMO is crystalline with a 110 out-of-plane texture.

clearly identified in a columnar-structured PCMO. The high-resolution TEM (HRTEM) image in Figure 4.11 demonstrates a well-defined and unreacted oxide/metal interface between the PCMO layer and the Pt BE.

4.4.2 Statistics of RSD Resistance

Five series of PCMO-based RSDs with different geometry sizes and aspect ratios—1,000 nm × 2,000 nm, 500 nm × 1,000 nm, 300 nm × 600 nm, 200 nm × 400 nm, and 160 nm × 160 nm—were patterned and fabricated. The thickness of the PCMO layer in all RSDs was fixed at 30 nm. Figure 4.12 shows the distributions of the measured high resistance state of the RSDs with various sizes under a sensing current of 1 mA. The statistics of all RSD series, including the mean (m), standard deviation (s), m/s ratio, and 95% lower bound (LB) and upper bound (UB) of the RSD resistances are listed in Table 4.1. Because all of the RSD devices tested had the same thicknesses of PCMO and TE/BE

FIGURE 4.10
Cross-sectional TEM image of a PCMO-based RSD.

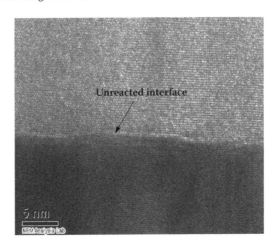

FIGURE 4.11
HRTEM image shows an unreacted PCMO/Pt interface.

layers, the resistance-area products (RAs) of these devices were expected to be the same. However, our measurement results in Table 4.1 demonstrate a large variation of the mean of RA (RA_{mean}) among devices of different sizes. The existence of defects such as grain boundaries is believed to cause such variations.

Figure 4.13 compares the measured RSD resistances with the direct scaling trend. Here we assume that in direct scaling, the same RA of 500 nm × 1,000 nm RSD is maintained as the device area scales down. Therefore, the device resistance (R) is reversely proportional to the device area (A). The discrepancy between the measured data and the predicted scaling trend presents the RA

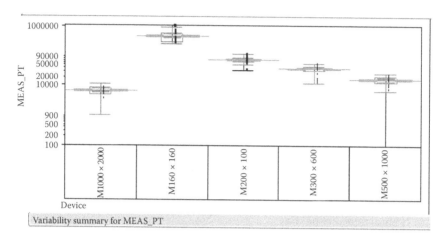

FIGURE 4.12

Distribution of the measured high-resistance state of PCMO-based RSDs with various sizes.

TABLE 4.1

Statistics of RSD High Resistance

Size (nm²)	μ (KW)	σ (KW)	μ/σ	95%LB (KΩ)	95%UB (KΩ)	RA$_{mean}$ (KΩ·mm²)
160 × 160	462.7	196.9	2.35	387.8	537.5	11.85
200 × 400	74.66	16.38	4.56	68.54	80.78	5.97
300 × 600	37.12	10.77	3.45	33.10	41.14	6.68
500 × 1,000	16.37	5.691	2.88	14.28	18.46	8.18
1,000 × 2,000	6.682	2.452	2.73	5.749	7.615	13.36

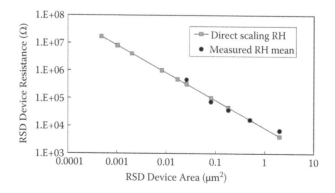

FIGURE 4.13

Measured high resistances of an RSD in comparison with the predicted scaling.

variations among devices of different sizes. Although the crystalline PCMO films are considered to be conductive, they cannot be simply treated as resistors.

In theory, large devices should have a tight resistance distribution because of the relatively small impact of the geometry variation on the actual device size. However, the m/s ratios of RSD resistances shown in Table 4.1 do not support this statement. We believe that this is due to the greater possibility of impurities and defects existing in larger devices than in the smaller devices.

In fact, the discrepancy between the theoretical prediction based on a bulky device and the actual properties of nanoscale devices was an important motivation for us to conduct this systematic study on the electrical properties of RSDs in scaled technologies.

4.4.3 Electrical Properties

As expected, bipolar resistance switching was observed in our fabricated PCMO-based RSDs. The measured DC sweep *I-V* hysteresis curve of an RSD with a size of 500 nm × 1,000 nm is shown in Figure 4.14. The RSD resistance switches from the high state to the low state at +1.6 V, whereas the device resistance switches from a low state to a high state at –2.2 V. We noticed that the DC voltage required to switch an RSD resistance does not show significant dependence upon the device size. For example, similar switching voltages in both resistance switching directions (about +1.6 and –2.2 V, respectively) were also observed for the 1,000 nm × 2,000 nm RSD, as shown in Figure 4.15. The quality of RSD degrades after a few cycles, which is demonstrated by the inconsistent I-V hysteresis curves shown in Figure 4.15.

Electric pulse–induced resistance (EPIR) bipolar switchings were also observed. Figure 4.10 shows the EPIR performance of a 500 nm × 1,000 nm PCMO device for 1,500 programming cycles. A positive electric pulse—that is, 2.5 V with 50 ms pulse width—was applied to the TE of the RSD and

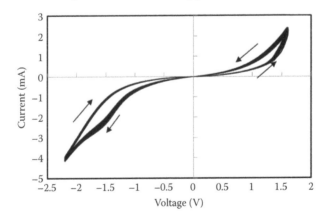

FIGURE 4.14
DC sweep *I-V* hysteresis curves of a 500 nm × 1,000 nm PCMO-based RSD.

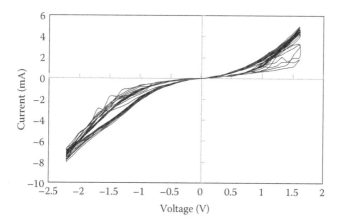

FIGURE 4.15
DC sweep *I-V* hysteresis curves of a 1,000 nm × 2,000 nm PCMO-based RSD.

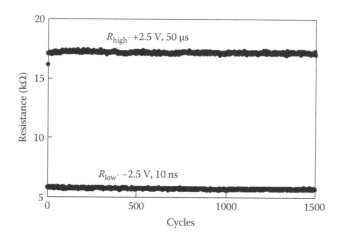

FIGURE 4.16
EPIR performance of a 500 nm × 1,000 nm PCMO-based RSD.

switched the RSD from a high-resistance state to a low-resistance state. On the other hand, a negative electric pulse—that is, −2.5 V with 10 ns pulse width—switched the RSD from a low-resistance state back to a high-resistance state. After each switching, the RSD resistance was measured under a sensing current of 1 mA.

Here R_{high} and R_{low} denote the resistance of RSD in a high-resistance state and low-resistance state, respectively. We define the EPIR ratio as $(R_{high} - R_{low})/R_{low}$, which is close to 200% in Figure 4.16. The means of the R_{high}s and R_{low}s at all programming cycles were 17.16 and 5.69 KW, respectively. Also, the resistive switching of the RSD shows excellent cycle-to-cycle stability: the standard deviations of R_{high} and R_{low} were only 81 and 44 W, respectively. We did not observe any significant degradation of the two resistance states after

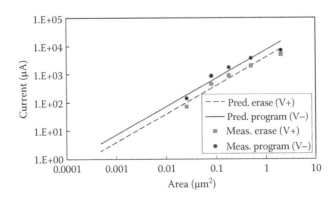

FIGURE 4.17
Measured program and erase currents in comparison with the predicted scaling.

1,500 programming cycles. This PCMO-based RSD shows good endurance stability with a smaller switching voltage close to ±2.5 V. The reduction of switching voltage was mainly due to the decrease in PCMO layer thickness.

Figure 4.17 shows the scaling trend of progamming/erasing (P/E) currents for various RSD sizes. Because RA stays the same and the switching voltage is insensitive to RSD size, the required P/E current decreases as RSD size shrinks down. When the RSD size equals 160 nm × 160 nm, the program and erase currents are reduced to 188 and 102 mA, respectively.

A high device yield was achieved. Figure 4.18 shows a testing results map of total 49,500 nm × 1,000 nm RSDs tested in one wafer (gray blocks indicate the devices that were not tested). DC resistive switching was obtained in 47 devices, which can be approximately translated to a 95% bit yield. Highly consistent electrical properties of RSDs were demonstrated: almost identical DC *I-V* hysteresis curves were obtained for all 47 devices, and even their locations were uniformly distributed over the wafer.

4.5 R-RAM Array Structure

4.5.1 Conventional Crossbar Array

The most popular R-RAM design is the crossbar array introduced in Section 4.3.1. Figure 4.19 shows a simple memory island built with a crossbar array. Figure 4.20 shows a scanning electron microscope (SEM) image of a crossbar memory island made by the University of Michigan [17]. Two sets of orthogonal metal wires were used for access control and a R-RAM cell sits at each cross point. Note that a R-RAM cell includes both storage material and the switching device. A crossbar array can potentially obtain the smallest memory cell area of $4F^2$.

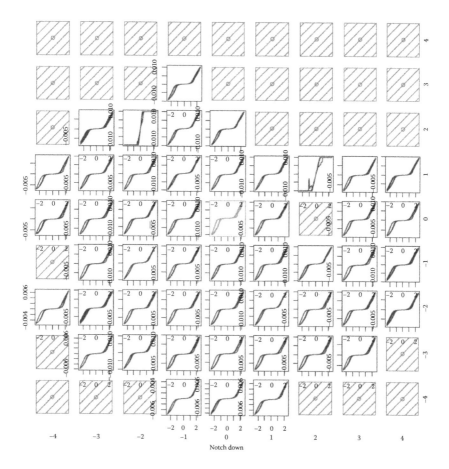

FIGURE 4.18
DC sweep *I-V* hysteresis curve of a whole wafer. Gray blocks represent the devices that were not tested. Resistive switching was observed in 47 of a total of 49 tested devices.

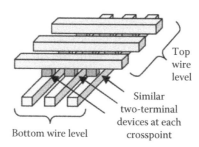

FIGURE 4.19
Simple crossbar structure of memory island [9].

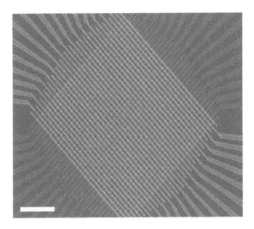

FIGURE 4.20
SEM of crossbar memory island [17].

Accessing a memory cell in a crossbar array can be simply conducted by providing the required biasing condition to the target cell. The terminals of the unselected cells can keep floating or be tied to ground. To read the stored data in a cell, we can supply a small read voltage to its WL and detect the current on its BL. This can be extended to multiple cells on the same WL. All or some of the selected cells can be read out simultaneously from the BLs. The bandwidth depends on the array size as well as the capacity of the readout circuit. A similar scheme can apply to a write scheme, and multiple cells on the same WL can be written together. However, programming and erasing operations have to be separated into two steps because they require different biasing conditions.

The design considerations of the conventional crossbar array have been thoroughly discussed in Liang and Wong [3]. Due to the sneak path, increasing the array size can dramatically reduce the sense margin in read operations as shown in Figure 4.21 [3]. Here, the R-RAM material has R_{on} = 5 KΩ and R_{off} = 1 MΩ. In an ideal design without considering the interconnect resistance, the sense margin is degraded by 30% in a 32 × 32 array. When increasing the array size to 64 × 64, the degradation is greater than 50%. If considering the interconnect resistance, the situation becomes even more severe—the data in a 64 × 64 array can barely be read out. The impact of the sneak path can be suppressed when increasing R_{on} of the R-RAM device. Therefore, the sensing margin of a crossbar memory is improved. In Figure 4.22, a trend of the minimal required R_{on} is given for different array capacities.

4.5.2 Complementary Crossbar Structure

It is hard to expand the conventional crossbar array without a switching device to a large array because the sensing margin degrades due to the conducting current through the sneak path. Recently, a complementary crossbar structure was proposed to solve this issue [18]. As shown in Figure 4.23, two

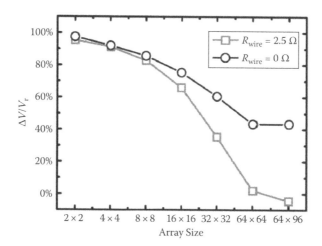

FIGURE 4.21
Impact of interconnection resistance [3].

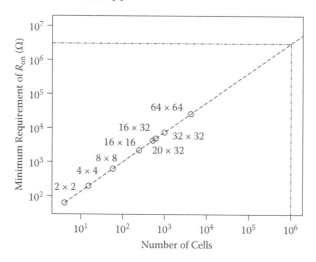

FIGURE 4.22
Minimum requirement of R_{on} in a conventional crossbar memory island [3].

identical R-RAM devices are stacked back to back to form a complementary resistive switch (CRS). One of the R-RAM devices was a switch cell and the other was a memory cell.

In this design, the data stored in the memory and the switch cells are complementary. For example, if the memory cell stores '0', the data in the corresponding switch cell must be '1.' Hence, the overall resistance of an unselected CRS remains high. Figure 4.24 shows the *I-V* characteristic of the complementary structure R-RAM cell [18]. For the high-resistance state of all of the unselected cells, the current conducted through the sneak path can be

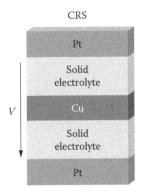

FIGURE 4.23
Complementary R-RAM cell [18].

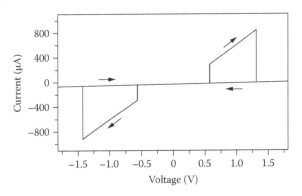

FIGURE 4.24
I-V curve of a complementary structure [18].

suppressed and the memory array size can be significantly improved, as we can see in Figure 4.25.

Accordingly, multiple steps are required to complete one device access. For instance, to read out the stored data in one CRS, we need to (1) turn on the switching device, (2) read out the memory cell, and (3) turn off the switching cell. The complementary structure has drawbacks of poor endurance and slow operation speed due to the multiple-step operation.

4.6 3D Stacking R-RAM

Three-dimensional stacking that builds up multiple memory layers vertically is an efficient way to improve density. For a 3D stacking R-RAM with *N* memory layers, the memory array density becomes *N* times that of a

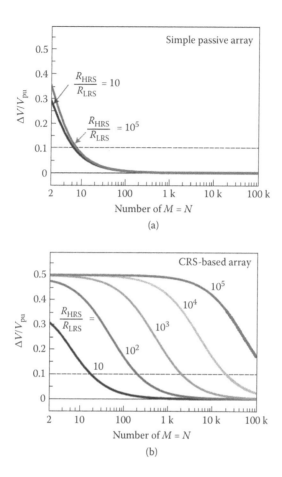

FIGURE 4.25
Read voltage margin: (a) normalized read voltage margin for a crossbar array with a stored worst-case pattern and (b) normalized read voltage margin for a crossbar array consisting of a CRS and the same stored bit pattern [18].

single-layer memory. To design a 3D stacking R-RAM structure, we need to consider all related components, including R-RAM cell characteristics, peripheral circuit design and placement, as well as the device stacking process limitations [20].

Figure 4.26 shows a conventional 3D stacking memory island in which memory layers are simply stacked on top of one another. Each memory layer has its own set of storage elements and interconnects. An isolation layer is inserted between two neighboring layers to prevent signal interference. Spin-on-glass (SOG) technology with methylsilsesquioxane (MSQ) materials can be used to form isolation layers [4, 21]. However, from a process development point of view there are some critical difficulties, including device degradation due to thermal processing, misalignment of vias due to SOG,

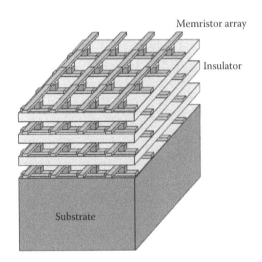

FIGURE 4.26
Conventional 3D stacking memory island.

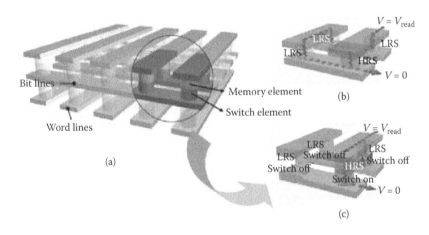

FIGURE 4.27
Stacking structure of a unipolar R-RAM crossbar array [22].

poor adhesion of SOG material, and heat accumulation because of the low conductivity of the isolation material.

4.6.1 3D Unipolar R-RAM Design

Figure 4.27 shows a stacking structure of unipolar R-RAM crossbar array developed by Samsung [22]. The single-cell structure is 1D1R. The detailed composition of the memory cell and switching cell in the stacking structure can be seen in Figure 4.28. There are two deposition orders—the forward

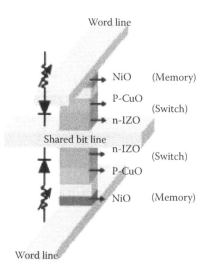

Word line

NiO (Memory)

P-CuO
(Switch)

n-IZO

Shared bit line

n-IZO
(Switch)

P-CuO

NiO (Memory)

Word line

FIGURE 4.28
Composition of the stacking structure of a forward deposition device and reverse deposition device [22].

deposition and reverse deposition—so that the adjacent memory layers share the same BLs. However, WLs cannot be shared. So when building up more than two layers, isolation is still required between two groups of WLs. Because a diode is used as the switching device, it can successfully cut off the reverse biasing from other layers. The read and write operations are similar to those of a one-layer crossbar array. In this design, memory cells in multiple layers can be accessed simultaneously.

4.6.2 3D Bipolar R-RAM Design

The 3D design with bipolar R-RAM cells can be achieved by building up multiple memory layers with an insulator in between. In order to avoid malfunctions caused by signal interference when simultaneously accessing multiple memory layers, the insulator layer cannot be omitted because the biasing conditions in both directions are needed. In such a design, each memory layer is independent of the other layers. Therefore, the memory access is exactly the same as the operation in a single-layer crossbar array.

Meier et al. proposed a two-layer stacking structure with resistively switching Ag-doped spin-on glass [23]. Figure 4.29 shows SEM images of various nanocrossbar structures. Arrays and word structures with feature sizes of 200 and 100 nm were realized. Another successful example is the four-layer bipolar 3D stacking structure proposed by Unity Corporation [24], which is illustrated in Figure 4.30. CMOx R-RAM devices with a noble electrode were used to construct the memory island and program/erase operations were well controlled. Their special two-layer structure with CMO and IMO, which

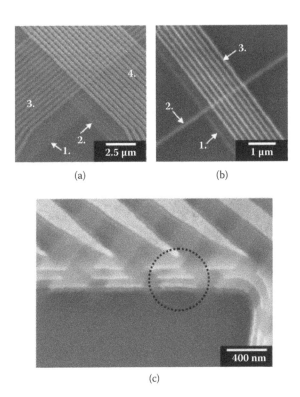

(a)   (b)

(c)

FIGURE 4.29
SEM images of multilayer crossbar memory structures with a 45-degree angle of view: (a) stacked 16-bit arrays with a half pitch of 200 nm; (b) 8-bit word structure with a half pitch of 100 nm; and (c) cross section [23].

represent conductor and insulator, respectively, prevent the occurrence of unnecessary overwrite.

4.6.3 3D High-Density Interleaved Memory Design

Figure 4.31 illustrates the 3D high-density interleaved memory (3D-HIM) design for a bipolar R-RAM cell. For simplicity, only six memory layers are demonstrated. A crossbar array is utilized in each layer. The basic design concept of 3D-HIM is to employ complementary material stack structures— that is, a regular memory stack and one with a reversed deposition order—to the memory cells in neighboring layers. For instance, all of the memory cells of layer 1 in Figure 4.31 use the regular deposition process (dark gray pillars), and those of layer 2 are made by reversing the deposition sequence (light gray pillars). The two types of memory stacks are applied to the odd and even layers alternatively. In this design, memory devices and metal wires form a memory island without isolation layers. Any two adjacent memory cells at the same *x-y* location are connected back to back and hence share the

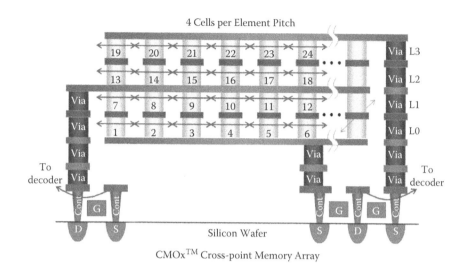

4 Cells per Element Pitch

FIGURE 4.30
Four-layer 3D structure with bipolar CMOx R-RAM devices [24].

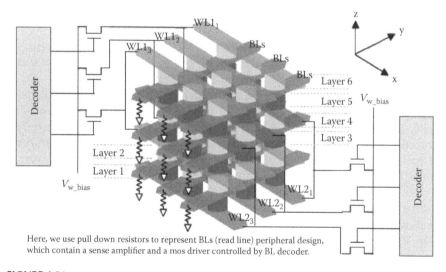

Here, we use pull down resistors to represent BLs (read line) peripheral design, which contain a sense amplifier and a mos driver controlled by BL decoder.

FIGURE 4.31
3D high-density interleaved memory design for bipolar R-RAM cell.

metal wire in between. The following terms are defined to help understand the proposed structure and corresponding operations:

BLs: A set of metal wires connected to TEs of R-RAM devices, which route along the y-axis as shown in Figure 4.31.

WLs: A set of metal wires connected to BEs of R-RAM devices, which route along the x-axis. There are two sets of WLs, called WL1 and WL2.

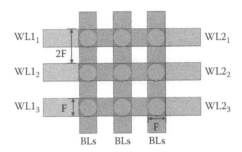

FIGURE 4.32
Layout of 3D high-density interleaved memory design.

WL1s and WL2s: We number the WL layer at the bottom of the 3D-HIM structure as '0' and continue counting the other WL layers from bottom to top. We define WL1s (WL2s) as those WL layers with odd (even) numbers.

WL1$_i$GC and WL2$_j$GC: We call the group of memory cells connected to a given WL1$_i$ or WL2$_j$ WL1$_i$GC (WL1$_i$ group cells) or WL2$_j$GC (WL2$_j$ group cells), respectively.

In all, three sets of control signals—that is, BL, WL1, and WL2—are utilized. Each is responsible for related operations in the memory layers above and below it.

Figure 4.32 illustrates a top view of the layout of a 3D-HIM. The cell area is $A_{3D-HIM} = 4F^2$, which is the same as the cell size of the conventional crossbar array ($A_{conv} = 4F^2$). Note that for a 3D memory, the density is determined not only by the single memory cell area but by the the allowable number of memory layers. By sharing BEs and TEs between neighboring layers, a 3D-HIM can reduce the overall number of conduction layers and remove isolation layers. For a given height of a 3D structure, which usually is a major limitation in the fabrication process, more memory layers can be stacked up vertically. Thus, the memory capacity is increased.

In 3D-HIM, there are two sets of group cells—WL1$_i$GC and WL2$_j$GC. Only one can be accessed at a time during read or write operations. We called this a *bigroup operation scheme*. This scheme has several advantages: (1) It increases throughput by simultaneously accessing multiple memory cells within either WL1$_i$GC or WL2$_j$GC. (2) The unselected groups can be biased to ground and taken as the signal isolators. Thus, we can avoid the unexpected overwriting caused by the write operations on different memory layers. (3) The BLs are shared by the R-RAM layers above and below the BLs. The peripheral circuitry connected to the BLs is also shared by two R-RAM layers to reduce area cost. Furthermore, WL1 and WL2 can be driven from opposite sides of the memory island as shown in Figure 4.31 to distribute the layouts of peripheral circuitry.

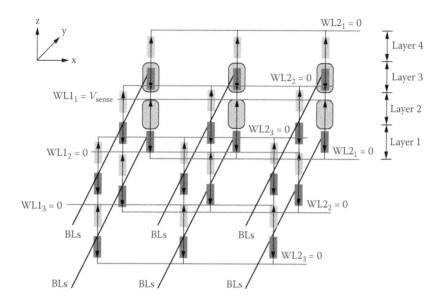

FIGURE 4.33
Selected WL₁GC in read operation.

4.6.3.1 Read Operation

To read out the stored data in an R-RAM cell, we provide a sense voltage (V_{sense}) to the corresponding WL and measure the current through the cell. To prevent unexpected overwriting, V_{sense} should be much smaller than the threshold voltage of the R-RAM device. A sense amplifier is connected to the BL and shared by two groups' cells WL1$_i$GC and WL2$_i$GC. Based on the bigroup operation scheme, only one group's cells can be sensed out at one time.

Figure 4.33 shows an example of reading out the cells in WL1$_1$GC. WL1$_1$ is raised to V_{sense} and all other WL1$_i$ are tied to 0 V. To prevent the disturbance from/to WL2 groups, all WL2s are forced to 0 V. Similarly, the read operation of WL2$_j$GC on the *x-y* plane can be accessed simultaneously (which is omitted here due to space restrictions).

An active load (R_{sense}) is used at the end of a BL to transfer current through the memory device to the input voltage of a sense amplifier $V_{R\text{-sense}}$. To simplify evaluation of the read operation in this work, we apply a 100 Ω resistance (R_{sense}) as the input resistance of the sense amplifier and define the SM by normalizing $V_{R\text{-sense}}$ with V_{sense}.

4.6.3.2 Write Operation

A bigroup operation scheme needs to be used in write operations to increase throughput and prevent unexpected overwriting. There are two possible

TABLE 4.2

Driving Conditions of Write Operations

Data	Cell Group	Driving Conditions
LRS	WL1$_j$GC	WL1: $-0.5V_{SET}$, BL: $0.5V_{SET}$
LRS	WL2$_j$GC	WL2: $-0.5V_{SET}$, BL: $0.5V_{SET}$
HRS	WL1$_j$GC	WL1: $-0.5V_{RESET}$, BL: $0.5V_{RESET}$
HRS	WL2$_j$GC	WL2: $-0.5V_{RESET}$, BL: $0.5V_{RESET}$

write procedures—set and reset. Like all bipolar R-RAM crossbar designs, these two procedures have to be separated because they require opposite driving polarities. In 3D-HIM, cells programmed at the same time must locate in the same group and have the same incoming value.

The driving conditions need to be carefully controlled to avoid unexpected overwriting caused by sneak paths and to minimize the total write current. The ideal bias voltages when performing set and reset are summarized in Table 4.2. All other WL1s, WL2s, and BLs that are not related to the write operation are forced to 0 V.

Figure 4.34 illustrates an example of WL1$_1$GC during a set operation. For illustration purposes, we assume that half of the cells in WL1$_1$GC are in the set procedure. WL1$_1$ are forced to $-0.5V_{SET}$, the BLs connected to the cells to be programmed are forced to $0.5V_{SET}$, and the unrelated control signals are set to 0 V. As shown in Figure 4.34, an unselected cell within WL1$_1$GC has only $0.5V_{SET}$ voltage drop across the cell, which is not large enough to change its content. The reset procedure is similar to the set procedure in the example. The corresponding bias voltages are summarized in Table 4.2. To save some space, we omitted the write operation of WL2$_j$GC on the *x-y* plane, which has similar setup requirements.

4.6.3.3 Impact of Data Patterns

The effectiveness of read and write operations in 3D-HIM depends on the memory data pattern. To investigate the impact of the data pattern, we divide all of the cells in a memory island into three categories: the target cell, the cells along the driving path (i.e., WL1 or WL2), and all remaining cells. Figure 4.35 shows an example of the WL1$_1$GC in the sensing operation. Four data patterns can be introduced—LL, LH, HH, and HL. Here, the first letter stands for the status of the target cell (L = LRS, H = HRS), and the second letter stands for the cells along driving path, either WL1s or WL2s.

Our work shows that the cells along the driving path and the rest of the cells dominate the SM rather than the target cell. The worst-case scenario occurs when the driving path cells and the rest of the cells are all in an LRS. In such a situation, the current conducted from the sneak paths and leakage

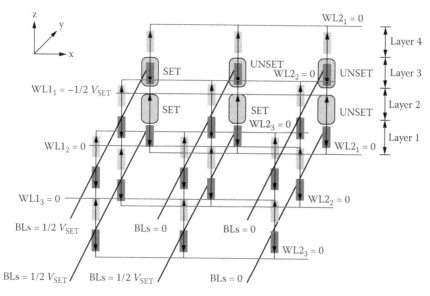

FIGURE 4.34
Selected WL$_1$GC in set operation.

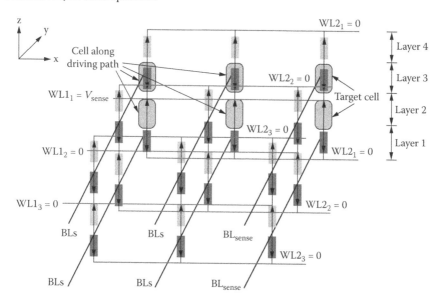

FIGURE 4.35
Three categories of cells used for data pattern analysis.

current are maximized. The corresponding data patterns are LL or HL. The impact of data patterns to the SM in our design is further discussed in the next section.

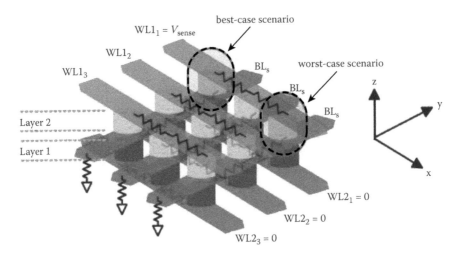

FIGURE 4.36
Worst-case and best-case scenario of the cell locations.

4.6.3.4 Impact of Cell Location

The physical location of a cell could affect its operating scenario. For example, Figure 4.36 illustrates a two-layer 3D-HIM in read operation of $WL1_1GC$. The driving current flows from the leftmost side of $WL1_1$ to the rightmost side of the array along the x-axis. Due to interconnect resistance, the real voltages applied on the cells along $WL1_1$ are not the same. The worst-case scenario occurs at the cell in the right corner because it goes through the longest path from the driver to the sense circuitry. In contrast, the cell located in the left corner is least affected by the interconnect resistance and hence it is the best-case scenario.

Figure 4.37 shows the SM difference between the worst-case scenario and the best scenario for cell locations with different data patterns in a four-layer 3D-HIM. As shown in the results, the impact of the array size on the LL pattern is much larger than the other patterns: the different cell locations difference could incur more than 10% variance in sensing margin. This is because the target cell in the LRS suffers from high interconnect resistance and the other cells on the sneak path in an LRS sink a large portion of the currents. To ease the impact of location difference, R-RAM of high R_{on}, which suffers less impact of current-resistance product (IR), is promising for large-scale arrays.

Figure 4.38 compares the sense margins of the conventional 3D bipolar R-RAM design and 3D-HIM under different memory configurations assuming that the array is under the worst-case scenario of cell location and data pattern. In general, 3D-HIM loses 10–20% in the sensing margin compared to conventional 3D R-RAM. This is because conventional 3D R-RAM inserts an isolation layer between any two memory layers; hence, the SM is determined only by one crossbar array. The control signals (e.g., WLs) in 3D-HIM

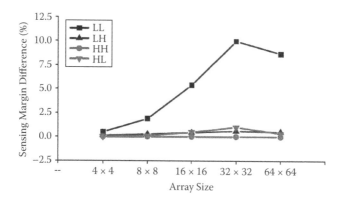

FIGURE 4.37
SM difference by the cell locations and data patterns.

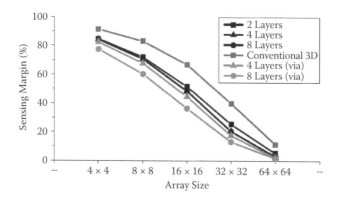

FIGURE 4.38
Sensing margin analysis of 3D-HIM.

have to drive twice the number of memory cells as the conventional design, which introduces more sneak paths. However, because of the interleaved design, stacking more layers in 3D-HIM only induces slight degradation of the sense margin. The SM decreases obviously as the array size increases. For example, in a four-layer 32×32 3D-HIM, the sense margin is about 20%. When the array size increases to 64×64, the SM is significantly reduced to 3%, which means that the status of memory cells is hard to detect. The comparison also shows that the via at small dimension could introduce a large resistance on the driving path, which incurs performance degradation in the upper layers.

The cell location and data pattern also affect the write operations. The worst-case scenario occurs at the same location and with the same data pattern as in the read operation. Due to space limitations, we only discuss the worst-case scenario and follow the explanation for read operation.

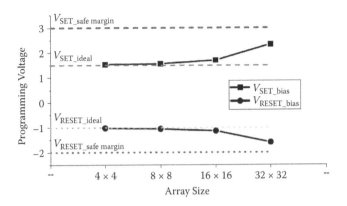

FIGURE 4.39
Proper programming voltage.

Enlarging the array size of a 3D-HIM increases the total IR in a driving path. To compensate for the impact of the increasing voltage drop on IR and properly program the target cells, a higher bias between WLs and BLs (V_{SET} or V_{RESET}) is required. The corresponding simulation for a four-layer 3D-HIM with various array sizes is shown in Figure 4.39.

The two dotted gray lines are the required set and reset voltages across an R-RAM cell, which are exactly $V_{SET\text{-}bias}$ and $V_{RESET\text{-}bias}$ in the ideal condition without IR. However, the impact of IR cannot be ignored in a real design and results in an increase in programming voltages as the array size increases, as demonstrated by the black curves. The dotted red lines constrain the safe margins of programming voltages, which double the range of the gray curves. If V_{SET} or V_{RESET} exceeds the safe margins, some unselected cells may be overwritten because their voltage drops are higher than the threshold. As a result, the proper programming voltage (black curves) and safe programming margins (red lines) confine the array size. Our simulation shows that the maximal allowable array size of a 3D-HIM is 32 × 32 to satisfy the constraints in write operations.

Compared to a conventional 3D stacking structure with an insulator layer, the 3D-HIM structure has advantages of increased density, immunity to thermal impact, and process simplicity.

4.7 Summary and Future

In this section, we introduce the R-RAM design from the perspective of the physical mechanism, single-cell design, array design, and three-dimensional structure.

In general, R-RAM denotes all random access memories that rely on the resistance differences to store data. Many materials or device stacks can demonstrate this characteristic for different physical and/or chemical mechanisms.

For the high density requirement, the crossbar array structure is dominant in R-RAM design, and 3D stacking can further improve the memory capacity within the given area constraint. However, increasing the size of the crossbar array is not easy because the sense margin in read operations decreases as the array size increases. Materials with high LRS can suppress the impact of sneak path current and the sense margin in read operations. In addition, novel design methodologies, such as complementary designs, were proposed.

R-RAM is a highly scalable device with inherent good characteristics that can be scaled down to the nanometer level. For instance, Ho et al. demonstrated 9-nm R-RAM with 10 MW LRS and R_{off}/R_{on} ~100 that can be used to build a R-RAM crossbar array larger than 64×64 [25]. The low programming current of 9-nm R-RAM, which is on the scale of milliampere (mA) or less, is compatible to the driving ability provided by the vertical BJT technology.

Retention time and endurance are two important parameters in R-RAM design. Retention time is the time that the device can maintain resistance in a tolerable range after a set/reset operation. Endurance is the limitation of write operation number of a R-RAM device. These parameters vary for different materials and switching mechanisms. For some R-RAM devices, greater than 10 years retention time [5, 25] and 10^6 rewrite numbers [5] have been successfully demonstrated.

References

[1] Kryder, M. H. and Kim, C. S. (2000). After hard drives—What comes next? *IEEE Transactions on Magnetics*, 45(10): 750–753.
[2] Baek, I. G., Kim, D. C., Lee, M. J., Kim, H.-J., Yim, E. K., Lee, M. S., Lee, J. E., Ahn, S. E., Seo, S., Lee, J. H., Park, J. C., Cha, Y.K., Park, S.O., Kim, H.S., Yoo, I. K., Chung, U-In, Moon, J. T., and Ryu, B. I. (2005). Multi-layer cross-point binary oxide resistive memory (OxRRAM) for post-NAND storage application. Paper presented at the IEEE International Electron Devices Meeting (IEDM), Washington, DC.
[3] Waser, R., Dittmann, R., Staikov, G., and Szot, K. (2009). Redox-based resistive switching memories—Nanoionic mechanisms, prospects, and challenges. *Advanced Materials*, 21(25–26): 2632–2663.
[4] Soni, R., Meier, M., Rudiger, A., Hollander, B., Kugeler, C., and Waser, R. (2009). Integration of Ge_xSe_{1-x} in crossbar arrays for non-volatile memory applications. *Microelectronic Engineering*, 86(4–6): 1054–1056.
[5] Jo, S., Kim, K., and Lu, W. (2009). High-density crossbar arrays based on a Si memristive system. *Nano Letters*, 9(2): 870–874.

[6] Li, H. and Chen, Y. (2009). An overview of non-volatile memory technology and the implication for tools and architectures. Paper presented at the IEEE Design, Automation & Test in Europe Conference & Exhibition, Nice, France.

[7] Liang, J. and Wong, H. (2010). Cross-point memory array without cell selectors—Device characteristics and data storage pattern dependencies. *IEEE Transactions on Electron Devices*, 57(10): 2531–2538.

[8] Chen, Y.-C., Chen, C. F., Chen, C. T., Yu, J. Y., Wu, S., Lung, S. L., Liu, R., and Lu, C.-Y. (2003). An access-transistor-free (0T/1R) non-volatile resistance random access memory (RRAM) using a novel threshold switching, self-rectifying chalcogenide device. Paper presented at the IEEE International Electron Devices Meeting (IEDM), Washington, DC.

[9] Lee, M.-J., Lee, C. B., Kim, S., Yin, H., Park, J., Ahn, S. E., Kang, B. S., Kim, K. H., Stefanovich, G., Song, I., Kim, S.W., Lee, J. H., Chung, S. J., Kim, Y. H., Lee, C. S., Park, J. B., Baek, I. G., Kim, C. J., and Park, Y. (2008). Stack friendly all-oxide 3D RRAM using GaInZnO peripheral TFT realized over glass substrates. Paper presented at the IEEE International Electron Devices Meeting (IEDM), San Francisco, CA.

[10] Wang, C.-H., Tsai, Y.-H., Lin, K.-C., Chang, M.-F., King, Y.-C., Lin, C.-J., Sheu, S.-S., Chen, Y.-S., Lee, H.-Y., Chen, F. T., and Tsai, M.-J. (2010). Three-dimensional $4F^2$ R-RAM Cell with CMOS logic compatible process. Paper presented at the IEEE International Electron Devices Meeting (IEDM), San Francisco, CA.

[11] Ahn, S.-E., Kang, B. S., Kim, K. H., Lee, M.-J., Lee, C. B., Stefanovich, G., Kim, C. J., and Park, Y. (2009). Stackable all-oxide-based nonvolatile memory with Al_2O_3 antifuse and p-CuO_x/n-$InZnO_x$ diode. *IEEE Electron Device Letters*, 30(5): 550–552.

[12] Yan, M. F. (1984). *Non-ohmic device using TiO_2*. U.S. Patent 4430255.

[13] Strukov, D. and Williams, R. (2009). Four-dimensional address topology for circuits with stacked multilayer crossbar arrays. *Proceedings of the National Academy of Sciences*, 106(48): 20155.

[14] Linn, E., Rosezin, R., Kugeler, C., and Waser, R. (2010). Complementary resistive switches for passive nanocrossbar memories. *Nature Materials*.

[15] Lu, Y. (2009). 3D technology based circuit and architecture design. Paper presented at the IEEE International Conference on Communications, Circuits and Systems, Milpitas, CA.

[16] Lewis, D. and Lee, H. (2009). Architectural evaluation of 3D stacked RRAM caches. Paper presented at the IEEE International Conference on 3D System Integration (3DIC), San Francisco, CA.

[17] Madayag, A. C. and Zhou, Z. (2001). Optimization of spin-on-glass process for multilevel metal interconnects. Paper presented at the Microelectronics Symposium, Richmond, VA.

[18] Lee, M.-J., Park, Y., Kang, B.-S., Ahn, S.-E., Lee, C., Kim, K., Xianyu, W., Stefanovich, G., Lee, J.-H., Chung, S.-J., Kim, Y.-H., Lee, C.-S., Park, J.-B., Baek, I.-G., and Yoo, I.-K. (2007). 2-Stack 1D-1R cross-point structure with oxide diodes as switch elements for high density resistance RAM applications. Paper presented at the IEEE International Electron Devices Meeting (IEDM), Washington, DC.

[19] Meier, M., Rosezin, R., Gilles, S., Rudiger, A., Kugeler, C., and Waser, R. (2009). A multilayer RRAM nanoarchitecture with resistively switching Ag-doped spin-on glass. Paper presented at the IEEE 10th International Conference on Ultimate Integration of Silicon (ULIS), Aachen, Germany.

[20] Bateman, B. (2011). Why CMOx™ cross-point memory arrays? Sunnyvale, CA: Unity Semiconductor Corporation.

[21] Ho, C. H., Hsu, C.-L., Chen, C.-C., Liu, J.-T., Wu, C.-S., Huang, C.-C., Hu, C., and Yang, F.-L. (2010). 9 nm Half-pitch functional resistive memory cell with <1 mA programming current using thermally oxidized sub-stoichiometric WOx film. Paper presented at the IEEE International Electron Devices Meeting (IEDM), San Francisco, CA.

5

Memristors

A memristor is an emerging nanodevice that can record the historical profile of the current or the voltage applied to itself [1, 2]. Thus, a memristor naturally integrates both computing and data storage capabilities simultaneously within itself. Combined with its ultrasmall footprint, the unique electrical properties of memristors create great potential for applications in very large-scale integration (VLSI) system designs, such as next-generation high-performance data massive storage [3] and neuromorphic computing systems (e.g., artificial neural networks, or ANNs) [4, 5], etc. Also, some special logic forms that are different from the traditional complementary metal-oxide semiconductor (CMOS) logic—for example, implication logic [6]—can be built around memristors.

In this chapter, we will first discuss memristor theory. Then we will introduce two important memristor types: metal-oxide and spintronic. Finally, we will discuss some applications of memristors, including memory and logic.

5.1 Memristor Theory

The three well-known basic passive circuit elements—resistor (R), capacitor (C), and inductor (L)—define the three mathematical relations connecting the pairings of four fundamental circuit variables—electric current (i), voltage (v), electrical charge (q), and magnetic flux (φ)—as

$$dV = R \cdot di \text{ (resistance)} \tag{5.1a}$$

$$dq = C \cdot dv \text{ (capacitance)} \tag{5.1b}$$

$$d(\varphi) = L \cdot di \text{ (inductance)} \tag{5.1c}$$

as shown in Figure 5.1. Based on the conceptual completeness of circuit theory, Professor Leon Chua in Berkeley predicted the fourth circuit element—the memristor—in 1971 [1]. A memristor defines the relationship between electrical charge and magnetic flux as:

$$d(\varphi) = M \cdot dq \text{ (Memristance)} \tag{5.1d}$$

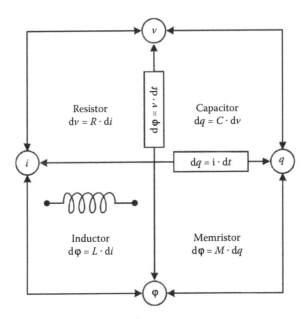

FIGURE 5.1
Four fundamental two-terminal circuit elements [2].

The unit of memristance is W—the same as that of a resistor—because

$$M = [d(j)/dt]/(dq/dt) = v/i \qquad (5.2)$$

When a memristor is a linear element, say, M is a constant, the memristor is nothing else but a constant resistor.

So what distinguishes a memristor from a resistor? In general, we cannot obtain a constitutive relation for a memristor in the v-i plane as we do for a resistor; the state of a memristor (memristance) is determined by the hysteretic behavior of the current through the device. Instead, a memristor has a constitutive relation in the q-φ plane, as shown in Figure 5.2.

In a more general form, Professor Chua defined a current-controlled memristive system in 1976 as [7]:

$$v = M(w, i, t) \cdot i \qquad (5.3a)$$

$$dw/dt = f(w, i, t) \qquad (5.3b)$$

where w is a set of variables describing the internal state of the system. M and f are some explicit functions of time. When M is the function of charge q (or the integral of current i over time) through the device—that is, $M[q(t)]$—we call it a *charge-controlled memristor*.

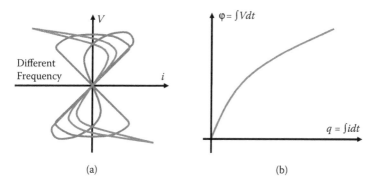

FIGURE 5.2
Memristor properties on the (a) v-i plane and the (b) φ-q plane.

On April 30, 2008, HP Lab announced the development of the first charge-controlled memristor [2]. The memristive effect was achieved in a thin-film structure of titanium dioxide (TiO_2) by moving the doping front along the device. Since then, many different types of memristors have been invented, which we will introduce in the next section.

5.2 Metal-Oxide Memristor

The first realization of memristors by HP can be categorized as the metal-oxide type, which is a metal–insulator–metal structure similar to the one adopted in resistive memory technology. Among many candidates, TiO_2 is probably the earliest and also the most popular.

5.2.1 Titanium Dioxide as Memristor

Figure 5.3 shows the equivalent model of a TiO_2 memristor with a Pt/TiO_2/Pt thin-film structure. Two metal wires (Pt) construct the top and bottom electrodes, and a thick titanium dioxide film is sandwiched in between. The stoichiometric TiO_2 with an exact 2:1 ratio of oxygen to titanium has a natural state as an insulator. However, if the titanium dioxide is lacking a small amount of oxygen, its conductivity becomes relatively high. We call this *oxygen-deficient titanium dioxide* (TiO_{2-x}).

The memristive function is achieved by moving the doping front: A positive voltage applied on the electrode connected to the doped area can drive the oxygen vacancies into the pure TiO_2 part and therefore lower the overall resistance continuously. Conversely, a negative voltage applied on the top electrode can push the dopants back to the TiO_{2-x} part and increase the overall resistance accordingly. We use R_{ON} (R_{OFF}) to denote the total resistance when a TiO_2 thin-film memristor is fully doped (undoped).

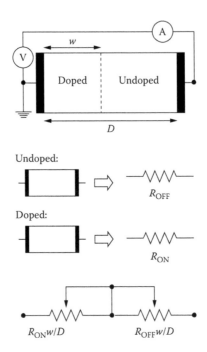

FIGURE 5.3
Coupled variable-resistor model for a memristor [2].

Figure 5.3 also illustrates a 1D coupled variable resistor model for TiO_2 thin-film memristors, where two resistors are connected in series. The overall resistance can be expressed as:

$$M(a) = R_{ON} \cdot \alpha + R_{OFF}(1-\alpha) \tag{5.4}$$

Here $\alpha = w/D$ $(0 < a < 1)$ is the relative doping front position, which is the ratio of doping front position (w) over the total thickness of the TiO_2 thin film (D). If we ignore the geometry variations and the nonuniformity of device parameters such as doping profile, the velocity of doping front movement $v(t)$, which is driven by the voltage applied across the memristor $V(t)$, can be expressed as:

$$v(t) = \frac{d\alpha(t)}{dt} = \mu_v \cdot \frac{R_L}{D^2} \cdot \frac{V(t)}{M(\alpha)} \tag{5.5}$$

where μ_v is the equivalent mobility of dopants and $M(\alpha)$ is the total memristance when the relative doping front position is α. Figure 5.4 depicts a photo of a memristor array demonstrated by HP.

Figure 5.5 shows the simulation result and the experimental result of a TiO_2 memristor under a sinusoidal voltage excitation of ±1 V. The gray line in

FIGURE 5.4
Photos of memristor array [32].

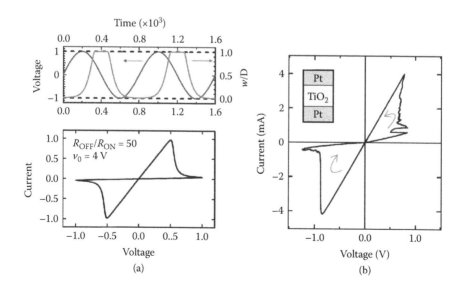

FIGURE 5.5
i-v Curve of TiO$_2$ memristor device: (a) simulation result and (b) measurement result [2].

Figure 5.5(a) shows the relative doping position within the device (α). The flat portion when the applied voltage equals 1 V denotes the situation that the doping front hit the boundary of the device. Further increasing the applied voltage does not change the total memristance (resistance) of the memristor. Figure 5.5(b) shows the experimental results from the real device.

5.2.2 Other Oxide-Based Memristor Options

Many metal-oxide structures also demonstrate that the voltage/current-driven resistance changes, though the resistance switch does not necessarily show the continuous changes. For example, Figure 5.6(a) shows the structure of a ZnO nanocrystal memristor [8]. The ZnO nanocrystal with 100 nm thickness is deposited on an indium–tin oxide (ITO) substrate and an aluminum (Al) contact is patterned on the top. A voltage of ±2 is applied to realize bipolar switching, as shown in Figure 5.6(b). Resistance state changes occur when the applied voltage is sufficiently high. However, continuous state changes, which mean that the resistance of the device can stop at any value, are normally hard to implement in such a structure.

After the TiO_2 memristor was invented, many memristor materials/structures based on various physical mechanisms were found or rediscovered, including spintronic devices [9, 10], polymeric thin film [11, 12], MgO-based magnetic tunnel junctions (MTJs) [13, 14], AlAs/GaAs/AlAs quantum-well diodes [15], ion conductor chalcogenide [16], etc. In the next section, we will introduce spintronic memristors, which were developed based on magnetic technology.

5.3 Spintronic Memristor

5.3.1 Spintronic Device as Memristor

In X. Wang et al. [9], three different types of magnetic memristors were proposed. Among these, the one based on the magnetic domain wall motion received the most attention from the solid-state device society. Figure 5.7 shows its structure and the simplified equivalent circuit model of the spintronic memristor. The spintronic memristor consists of a long spin-valve strip that includes two ferromagnetic layers: a reference layer and a free layer. The magnetization direction of the reference layer is fixed by coupling to a pinned magnetic layer. The free layer is divided by a domain wall into two segments that have opposite magnetization directions from each other. The resistance per unit length of each segment is determined by the relative magnetic directions of the free layer and reference layer: when the magnetization direction of the free layer in a segment is parallel (antiparallel) to the reference layer, the resistance per unit length of the segment is low (high).

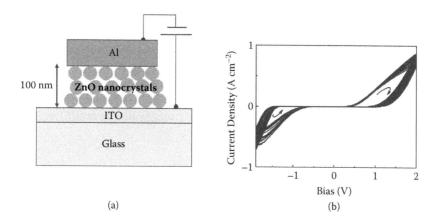

FIGURE 5.6
(a) ZnO-based memristor structure and (b) its experimental result [8].

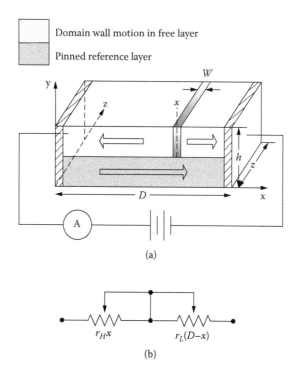

FIGURE 5.7
Spintronic memristor based on magnetic domain wall motion: (a) structure and (b) equivalent circuit.

Fabrication of the spintronic memristor could be relatively easier than with other types of memristors. In fact, the required fabrication technology is similar to the one used in the manufacture of the spin valve-based giant magnetoresistance (GMR) head of a hard disk drive [17].

We use r_L and r_H to denote the value of the resistance per unit length when the segment of a spin-valve strip is in a low- or high-resistance state, respectively. By ignoring the contribution of the domain wall, the memristance of a spintronic memristor can be calculated as:

$$M(x)=\left[r_H \cdot x + r_L \cdot (D-x)\right] \tag{5.6}$$

where x is the position of the domain wall. D is the length of the device, as shown in Figure 5.7(a). Normally the domain wall velocity v is proportional to the current density J [17]; that is,

$$v = \frac{dx}{dt} = \Gamma_v \cdot J = \frac{\Gamma_v}{h \cdot z} \cdot \frac{dq}{dt} \tag{5.7}$$

where the memristance of spintronic memristor M can be simply expressed as a function of the charge through the device:

$$M(q)=\left[r_L \cdot D + (r_H - r_L)\Gamma q(t)\right] \tag{5.8}$$

Here h and z are the thickness and width of the spin-valve strip, respectively. Γ_v is the domain wall velocity coefficient, which is related to the device structure and material property.

Usually the thickness of spin-valve strip h is determined by the fabrication technology. After the device structure and materials are determined, the upper bound and the lower bound of memristance can be adjusted by changing the geometry of the device; that is, length D and width z (or the cross-sectional area of the spin-valve strip). The variations of the device geometry, however, will incur variability of the memristance in the same way.

As is different from the motion of a doping front in the solid-state memristor, the domain wall movement in the spintronic memristor occurs only when the applied current density (J) is above the critical current density (J_{cr}) [18–23]. In other words, $G_v = 0$ when $J < J_{cr}$ in Eq. (5.7). As we will show later, J_{cr} is determined by many factors and varies from device to device.

When reading the memristance of a spintronic memristor, a small sensing current (I_{rd}) can be applied to the device. The value of memristance is read out by measuring the voltage drop across the memristor. As long as the read current density J_{rd} is below J_{cr}, the state of the spintronic memristor will not be disturbed.

5.3.2 Compact Model of Spintronic Memristor

5.3.2.1 *Model of Spintronic Memristor*

Memristance Dependence on Domain Wall Position x

By considering the contribution of the domain wall, the memristance of a spintronic memristor can be expressed as a function of domain wall position x as:

$$M(x) = \left[r_H \cdot (x - w/2) + R_{dw} + r_L \cdot (D - x - w/2) \right] \tag{5.9}$$

where R_{dw} is the resistance of domain wall and w is the width of domain wall. Here we assume that the domain wall position x is at the middle of the domain wall.

By assuming that the resistance per unit length of the thin film strip changes linearly from r_L to r_H within the domain wall (as shown in Figure 5.8), the memristance of the device can be calculated by:

$$M(x) = \left[r_H \cdot (x - w/2) + (r_H + r_L) \cdot w/2 + r_L \cdot (D - x - w/2) \right]$$
$$= \left[r_H \cdot x + r_L \cdot (D - x) \right] \tag{5.10}$$

for $0 < x < D$. Here $M(x)$ does not depend on the width of the domain wall. We note that such an assumption of the domain wall resistance is close to the physical phenomena and incurs a marginal error in the calculation of memristance [19–23].

Domain Wall Position x

The domain wall position x can be calculated by the integral of the domain wall velocity v over time t as:

$$x = \int_0^t v \, dt \tag{5.11}$$

The domain wall velocity v is proportional to the current density J as long as $J > J_{cr}$ [18–23]:

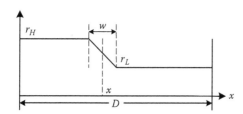

FIGURE 5.8
Profile of resistance per unit length along the device.

$$v = \begin{cases} \dfrac{JP u_B}{e M_s} & J \geq J_{cr} \\ 0 & J < J_{cr} \end{cases} \qquad (5.12)$$

Here P is polarization efficiency, u_B is Bohr magneton, e is an elementary charge, and M_s is magnetization saturation. The domain wall velocity coefficient Γ_v equals $P u_B / e M_s$.

By combining Eqs. (5.11) and (5.12), the domain wall position x can be calculated as:

$$x = \Gamma_v \int_0^t J_{eff} dt \left(J_{eff} = \begin{cases} J & J \geq J_{cr} \\ 0 & J < J_{cr} \end{cases} \right) \qquad (5.13)$$

Here J_{eff} is defined as the effective current density in the spintronic memristor.

Current Density J

The current density J in a spintronic memristor can be calculated as:

$$J = \dfrac{V}{M(x) \cdot h \cdot z} \qquad (5.14)$$

where V is the voltage applied to the spintronic memristor; h and z are the thickness and width of the spin-valve strip, respectively.

5.3.2.2 Electrical Property of Spintronic Memristor

Sheet Resistance Bounds

A memristor is a two-terminal device that can be modeled as a time-varying resistor in a circuit. The upper and lower bounds of sheet resistance are two important electrical properties of a spintronic memristor.

The square sheet resistance of the thin-film strip in the spintronic memristor is determined by the relative magnetic directions of the free layer and reference layer. We denote the square sheet resistances of the spintronic memristor when the magnetic directions of the two ferromagnetic layers are antiparallel or parallel as Re_H and Re_L, respectively.

A common parameter to represent the difference between Re_H and Re_L is the GMR ratio, which is defined as:

$$GMR = \dfrac{Re_H - Re_L}{Re_L} \qquad (5.15)$$

Low square sheet resistance Re_L is determined by the resistivity (ρ) of the thin-film strip in the low-resistance state and the thickness (h) as:

$$\mathrm{Re}_L = \frac{\rho}{h} \tag{5.16}$$

ρ can be considered as an intrinsic electrical parameter that is determined by the device structure and material property only.

The high and low resistance per unit length of the spintronic device r_H and r_L can be then calculated by:

$$r_L = \mathrm{Re}_L / z$$
$$r_H = \mathrm{Re}_L (1 + GMR)/z \tag{5.17}$$

Domain Wall Velocity Coefficient Γ_v

The domain wall velocity coefficient $\Gamma_v = P\mu_B/eM_s$ describes how fast the domain wall can move when applying a certain current density on the memristor and, consequently, the speed of memristance change. Instead of deriving Γ_v after measuring the magnetic parameters P, μ_B, and M_s, we directly measure Γ_v, which can be roughly considered as a normal distribution.

Critical Current Density J_{cr}

The critical current density J_{cr} can be theoretically calculated as:

$$J_{cr} = \frac{\alpha\gamma H_p eM_s}{P\mu_B}\sqrt{\frac{2A}{M_s H_k}} \tag{5.18}$$

Here, H_p is the hard anisotropy in the direction perpendicular to the thin-film plane (y direction), H_k is the easy anisotropy in the strip direction (x direction), A is an exchange parameter, α is a damping parameter, and γ is the gyromagnetic ratio.

It is well known that the theoretical calculation of J_{cr} cannot explain all experimental measurements. In magnetic device design, J_{cr} is usually considered an intrinsic electrical parameter that is directly obtained from the experimental calibration.

Domain Wall Modeling

Domain wall width w can be calculated by [18–23]:

$$w = \begin{cases} 2x & x < \sqrt{\dfrac{2A}{M_s H_k}}\big/2 \\[3ex] \sqrt{\dfrac{2A}{M_s H_k}} & \sqrt{\dfrac{2A}{M_s H_k}}\big/2 \leq x \leq D - \sqrt{\dfrac{2A}{M_s H_k}}\big/2 \\[3ex] 2(D-x) & x > D - \sqrt{\dfrac{2A}{M_s H_k}}\big/2 \end{cases} \tag{5.19}$$

TABLE 5.1

Constants and Parameters in Model

Physical Constants		
e	Elementary charge (C)	$1.602e^{-19}$
u_B	Bohr magneton (J·T^{-1})	$9.274e^{-24}$
Materials Parameters (Typical Value)		
H_p	Hard anisotropy (Oe)	5,000
H_k	Easy anisotropy (Oe)	100
M_s	Magnetization saturation (emu/cc)	1,010
A	Exchange parameter (J/m)	$1.8e^{-11}$
α	Damping parameter	0.002–0.1
P	Polarization efficiency	0.35
γ	Gyromagnetic ratio	$1.75e^7$
J_{cr}	Critical current density	5×10^7 A/cm²
Model Parameters		
D	Length (nm)	1,000
h	Thickness (Å)	70
z	Width (nm)	10
Re_L	Low sheet resistance (Ω/ð)	50 (when $h = 70$ Å)
GMR	Giant magnetoresistance ratio	12%

A typical value of w equals 59.7 nm, based on the material parameters shown in Table 5.1.

5.3.2.3 Model Summary

Table 5.1 summarizes the physical constants, the material parameters, and the model parameters involved in our model [33]. Figure 5.9 highlights the main steps and equations included in our proposed model. This model can be easily implemented by Verilog-A language to simulate the electric behavior of a two-terminal spintronic memristor in SPICE tool from Synosys Inc., CA.

5.3.3 I-V Curve and Frequency Response

We can use the proposed spintronic memristor model to simulate the behavior of the memristor device under different excitations.

5.3.3.1 M-q Curve

In a spintronic memristor, if the current density J is always above the critical current density J_{cr}, the relationship between memristance M and the charge through the device q ($= \int_0^t Jdt$) is unique, as shown in Figure 5.10.

Material parameters: J_{cr}, Γ_v
Model parameters: D, h, z, Re_L, GMR
High (low) resistance per unit length r_H (r_L):

$$r_L = Re_L/z$$

$$r_H = Re_L\left(1+GMR\right)/z$$

Domain-wall position x: x is the function of current density:

$$x = \Gamma_v \int_0^t J_{eff}dt \quad \left(J_{eff} = \begin{cases} J & J \geq J_{cr} \\ 0 & J < J_{cr} \end{cases}\right)$$

Current density J:

$$J = \frac{V}{M(x)\cdot h\cdot z}$$

Memristance M:

$$M = \left[r_H \cdot x + r_L \cdot (D-x)\right]$$

FIGURE 5.9
Model summary: main model components and key equations.

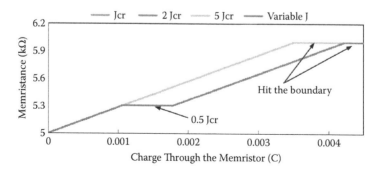

FIGURE 5.10
M-q curve of a spintronic memristor.

The magnitude of the current density—for example, J_{cr}, $2J_{cr}$, and $5J_{cr}$—does not affect the relationship between M and q. The material and the model parameters of the simulated spintronic memristor are shown in Table 5.1. When $J = J_{cr}$, domain wall velocity $v = 10$ m/s. It takes 100 ns for the domain wall to move from one end to the other end of a spintronic memristor with 1,000 nm length.

We note that the memristance of the spintronic memristor will not change if $J < J_{cr}$. In Figure 5.10, the curve "Variable J" shows the M-q relation when J_{cr} is applied in (0, 30 ns) and (70 ns, ~), and $0.5J_{cr}$ is applied in (30, 70 ns).

When $J \geq J_{cr}$, the memristance of the spintronic memristor keeps increasing until the position of the domain wall reaches the far end of the memristor; that is, $x = D$. However, the memristor cannot be simply modeled as a nonlinear current-controlled resistor of which the resistance is determined by the magnitude of the current through the device: the *I-V* curve of the spintronic memristor is not unique at all, as we shall show later on.

5.3.3.2 Sinusoidal Wave Excitation

To demonstrate the hysteretic profile of a memristor, we simulated the dynamic behavior of a spintronic memristor driven by a sinusoidal voltage of $V_m\sin(\omega t)$. The frequency of voltage $f = \omega/2p$ is set to 10 MHz (=1/100 ns). The simulation results of $V_m = 0.42$ and 1.05 V are shown in Figure 5.11.

When $V_m = 0.42$ V, the domain wall never reaches the far end boundary ($x = D$) of the memristor at $f = 10$ MHz. The flat top of the memristance curve is due to the insufficient current density; that is, $J < J_{cr}$. When $V_m = 1.05$ V, the flat top of the memristance curve is a result of the domain wall reaching the far end boundary of the memristor. The value of memristance M oscillates between 5 kW (=$r_L \cdot D$) and 6 kW (=$r_H \cdot D$).

The corresponding *I-V* curve of the simulated spintronic memristor is shown in Figure 5.12. We note that the shape of the *I-V* curve is dependent on the historical status of the applied voltage; for example, the frequency and the amplitude of voltage.

For comparison, we also simulated the *I-V* curve for the case where $f = 50$ MHz, $V_m = 1.05$ V as shown in Figure 5.12. It clearly shows that the *I-V* curve of the spintronic memristor is not unique and depends on the historical profile of the electrical excitations.

5.3.3.3 Biased Sinusoidal Wave Excitation

Another example is the dynamic behavior of a spintronic memristor with a biased sinusoidal voltage excitation. We set $f = 20$ MHz and $V_m = 0.42$ V. A positive bias (=0.05 V) is added for the first four cycles and then a negative bias (= −0.05 V) is added for the next four cycles, as shown in Figure 5.13.

In the first four cycles, under impact of both the positive bias and the sinusoidal voltage component, the memristance increases with some ripples.

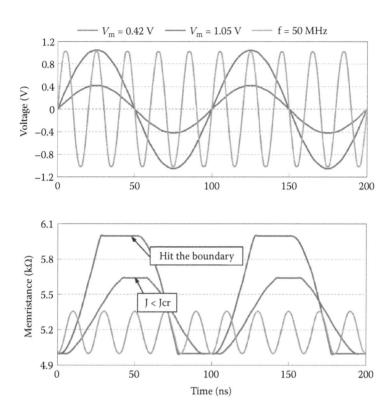

FIGURE 5.11
Spintronic memristor driven by a sinusoidal voltage.

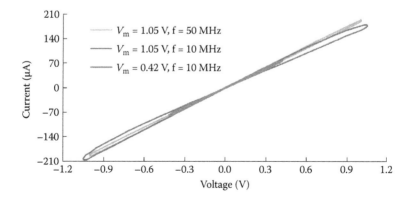

FIGURE 5.12
I-V curve of the spintronic memristor with sinusoidal voltage excitation.

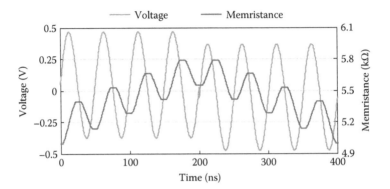

FIGURE 5.13
Memristance of a spintronic memristor driven by a biased sinusoidal voltage.

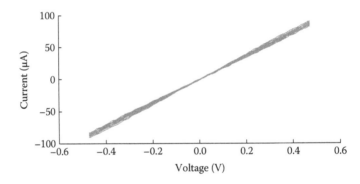

FIGURE 5.14
I-V curve of the spintronic memristor with a biased sinusoidal voltage excitation.

Similarly, in the next four cycles, the domain wall moves back to the near end of the device and memristance decreases to the original value.

The corresponding *I-V* curve is shown in Figure 5.14. Because of the accumulative effect of the voltage bias, the *I-V* curve of each cycle keeps shifting from that of the previous cycle.

5.4 Applications and Future Trends

5.4.1 Memristor Memory

Attractive properties of the memristor such as nonvolatility, high distinguishability, low power, scalability, and fast accessing speed make it a promising candidate for the next-generation high-density and high-performance memory technology [24, 25]. The resistance of a memristor device can be used to represent the value stored in the memory cell. The basic structure of a memristor-based memory is very similar to a typical memory array such as static random access memory (SRAM)-based and dynamic RAM (DRAM)-based memory, which consists of the word line decoder, memory array, sense amplifier, and output multiplexor (MUX). However, due to the special electronic characteristics of a memristor, several additional peripheral circuits, such as a pulse generator and read–write (R/W) selector, should be added to implement the write–read operations.

Because the resistance change of a memristor device is continuous, a multilevel cell (MLC) can also be easily implemented by programming the device resistance to the desired value. However, small resolutions between different resistance states of an MLC memristor device may limit the sense margin of the device.

In Niu et al. [26], a dual-element memory design was proposed based on memristor technology. It was observed that if we do not use the whole range of the resistance and only program a part of the memristor, the energy consumption can be saved because the resistance changes of a memristor are not linear, associated with the reduced write pulse width.

As shown in Figure 5.15, the time for programming the cell from R_{High} to $(R_{\text{High}} - R_{\text{Low}})/2$ is reduced to 75% of the original time. Here R_{High} and

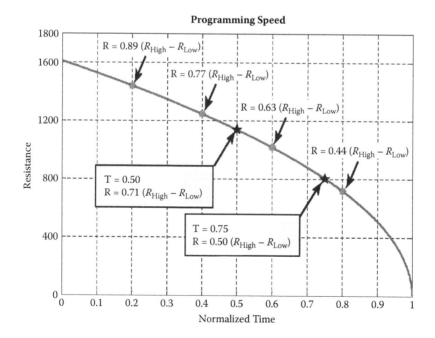

FIGURE 5.15
Programming time vs. resistance of a memristor [26].

R_{Low} denote the highest and the lowest resistance of a memristor device. On the other hand, if we program the cell from R_{Low}, the programming time is reduced to 25%. Intuitively, programming the low-resistance part of the memristor is faster than programming the high-resistance part. Therefore, we can take advantage of this kind of partial programming to optimize the memristor-based memory.

It is obvious that the resistance resolution of the memristor cell is reduced under such partial programming techniques. For a memristor-based memory cell, the read operation is realized by applying a small voltage across the cell and then sensing the current through the cell. Because the magnitude of the current is determined by the resistance of the memory cell, to ensure the reliability of the read operation, the current accessing the cell should be distinguishable enough.

To overcome the reduced resistance resolution, a dual-element memristor-based memory was proposed [26]. The basic idea of the dual-element memory cell is to use a pair of memristors to store the differential signals of the input data. Therefore, a large difference can be read by sensing both memristors. During the write operation, the differential signals are saved in the cells separately. One of the cells, called the positive cell, stores the data with the same polarity of the input data, and the other one, the negative cell, stores the complementary data. For example, in order to write logic '0' into the cell, the positive cell is written to high resistance. At the same time, the negative cell is written to low resistance.

The read scheme of the proposed differential memristor-based memory design is shown in Figure 5.16(b): during the read operation, instead of comparing the voltage to a reference resistance (in Figure 5.16(a)), the voltages from the two memristors are compared. Therefore, the voltage gap input to the sense amplifier is also increased.

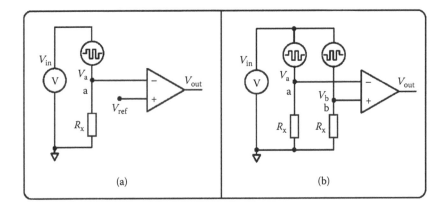

(a)   (b)

FIGURE 5.16
Sensing scheme for single/dual-element memristor-based memory cell: (a) single memristor cell sensing and (b) dual memristor cell sensing [26].

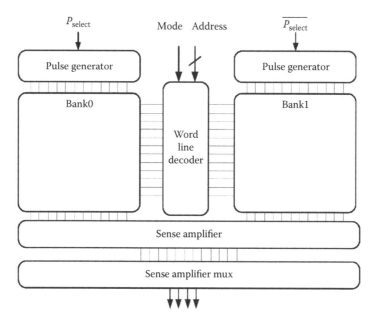

FIGURE 5.17
Overall structure of the memory array [26].

A dynamic memristor-based memory can be designed to easily switch between a high-capability, normal mode and a low-capability, low-power mode, with very small area and energy overhead compared to general memristor-based memory.

Figure 5.17 shows the overall structure of the proposed memory array with peripheral circuits. Generally, the memory structure is based on the design proposed in Ho et al. [24]. A metal-oxide semiconductor field-effect transistor (MOSFET) is used as the selector to reduce the leakage current through the memory cell. Several additional circuits are added to the design to ensure that the memory can switch between the two modes. The memory is divided into two banks, bank 0 and bank 1. The positive cells are located in bank 0 and the negative cells are located in bank 1. Each bank has its own pulse generator. In mode 0, two banks work independently, whereas in mode 1, two banks work together to realize the dual-element memristor-based memory, and each pair of the associated cells is located at the same positions in every bank.

The pulse generator provides different voltage levels to the memory cell. A mode bit is assigned to each pulse generator to choose the voltage level. If mode bit is logic '1,' indicating that the memory works in low power mode, the low voltage V_L is applied to memristor cells. If mode bit is logic '0,' the normal mode uses the high voltage V_H for the circuit.

The read operation is similar to the single-cell memristor memory. As shown in Figure 5.16, to extract the information from the cell, a voltage V_{in} is applied across the whole cell and the current is sensed. In Figure 5.18, if the

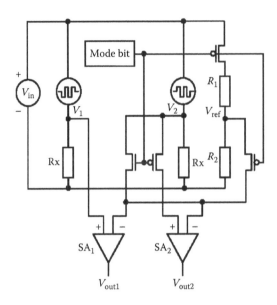

FIGURE 5.18
Circuit structure for dual-element memristor-based memory cell [26].

mode bit is set to 0, the circuit works as two independent memory cells. The voltages V_1 and V_{ref} are compared in sense amplifier 1 (SA$_1$) and V_2 and V_{ref} are compared in sense amplifier 2 (SA$_2$). However, if this bit is set to 1, the circuit works in low-power dual-element mode. The resistors used to generate reference voltage are useless and are isolated. Additionally, the voltages V_1 and V_2 are compared to extract the data.

In the read operation of a memristor device, the applied voltage will perturb the memristor state. If the mechanism is not well designed, the memristor state might be perturbed away beyond its safety margin, and a soft error will occur after repeated read when no recovery scheme is applied [24].

The proposed method for a read operation is in Figure 5.19(a). It has two stages: the convert stage and sense amplifier stage. Focusing on the convert stage as shown in Figure 5.19(b), let V_{in} be the input voltage and V_x be the voltage at node x. The purpose of adding a series resistor is to convert the memristor state into a voltage signal because the current through the memristor carries the memristor state information. The read pattern is illustrated in Figure 5.19(c). This read pattern will enforce zero net flux injection over a period of time to avoid altering the memristor state after read cycles.

5.4.2 Memristor Thermal Sensor

Some memristor devices can also be used as the sensing device; for example, as a thermal sensor [27]. In the spintronic memristor, the domain wall moves under the current-induced spin-torque excitation. For a ferromagnet

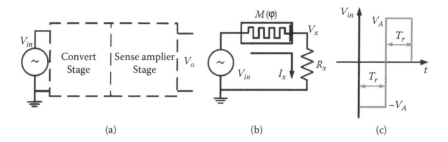

FIGURE 5.19
(a) Read operation stages; (b) convert stage circuit; and (c) read signal pattern [24].

with magnetization saturation M_s, exchange strength A, easy z-axis anisotropy H_k, and hard y-axis anisotropy H_p, the spin-torque-induced domain wall motion at a finite temperature is described through the stochastic Landau-Lifshitz-Gilbert equation [28] with a spin-torque term [29, 30]. Using rigid wall approximation [23, 31], the domain wall motion is expressed in terms of magnetization spherical angle θ, φ as: $\theta(x,t) = \theta_0[x - X(t)]$, $\phi(x,t) = \phi_0(t)$, where $\theta_0(x) = \arcos[\tanh(x/w)]$ is the function of domain wall shape with a wall thickness $w = \sqrt{2A / M_s H_k}$. $X(t)$ is the domain wall position. Domain wall speed is $dX(t)/dt$. The domain wall position $X(t)$ satisfies the following stochastic differential equations [23, 31]:

$$\frac{d\varphi}{dt} + \frac{\alpha}{w}\frac{dX}{dt} = \eta_\varphi$$

$$\frac{1}{w}\frac{dX}{dt} - \alpha\frac{d\varphi}{dt} = \omega_0 \sin(2\varphi) + \frac{v_s}{w} + \eta_X$$

(5.20)

where $\omega_0 = \gamma H_p/2$ (γ is the gyromagnetic ratio), α is the damping parameter, and $v_s = PJ\mu_B/eM_s$ is the spin-torque excitation strength. P is the polarization efficiency, μ_B is the Bohr magneton, and e is the elementary electron charge. $\eta_\varphi(t)$ and $\eta_X(t)$ are φ and X component thermal fluctuation fields.

Equation (5.20) shows that domain wall mobility at a finite temperature is in fact determined by both spin-torque excitation strength and thermal fluctuation magnitude or a function of the normalized spin-torque excitation strength and the normalized thermal fluctuation magnitude.

Here spin-torque excitation strength is proportional to current density. The normalized current density is defined as $PJ\mu_B/eM_s ww_0$. At zero temperature, the domain wall starts to move only when the current density is above the critical value, $PJ\mu_B/eM_s ww_0 = 1$. At a finite temperature, the domain wall can move even when the current density is below a critical value. The average domain wall velocity (normalized by v_s) as a function of the normalized current density and the normalized thermal fluctuation magnitude $(4k_B T/N\hbar\omega_0)$ can be obtained by solving stochastic differential Eq. (5.20).

Figure 5.20 shows the normalized domain wall velocity as a function of the normalized current density for different normalized thermal fluctuation magnitudes. Temperature-sensitive and -insensitive regions can be observed. Curves with kneeling shapes are around the critical current density, where the domain wall velocity is sensitive to thermal fluctuation magnitude.

For temperature sensing, a biasing voltage pulse with constant magnitude is applied to the spintronic memristor. The resistance difference before and after voltage pulse is measured. This resistance difference is calibrated to sense temperature magnitude. Figure 5.20 shows that domain wall mobility increases as temperature increases. When the domain wall moves to the low-resistance end, the total memristance drops continuously. Thus, after applying a constant magnitude voltage pulse, the device resistance drop is higher at higher temperatures.

The temperature-sensing memristor can operate at a region where its electric behavior is sensitive to temperature changes. This is achieved through a combination of temperature-dependent domain wall mobility and the positive feedback between resistance and driving strength in the memristor. Figure 5.20 shows the sensitive dependence of domain wall velocity upon temperature at the kneeling region around a critical current value. The positive feedback between resistance and driving strength is a unique property of the memristor: For a constant voltage pulse driving, higher temperature results in an increased domain wall moving distance. The increased domain wall moving distance results in a smaller resistance. The smaller resistance results in a higher driving current density under a constant voltage excitation, thus providing positive feedback to further increase domain wall distance. This positive feedback accelerates domain wall speed and reduces device resistance further for constant voltage pulse driving.

Figure 5.21 shows a temperature-sensing principle of a device example. We use a square-wave pulse with duration 80 ns and magnitude 0.3 V to excite the device. The device resistance drops as a function of time for different temperatures (300, 350, 400 K). Device resistance eventually flattens out when the domain wall settles to a fixed position after voltage pulse excitation. Solid curves in the figure are the resistance changes for the proposed spintronic memristor at different temperatures. Dashed lines are the resistance drop for a nonmemristive device without positive feedback between resistance and the integration of driving strength or under a fixed driving current strength. It can be seen that positive feedback between resistance and driving strength in a memristor significantly increases the temperature-sensing margin.

The material of the spintronic memristor temperature sensor is similar to the commercial recording head material. The sensing temperature range of our proposed device is between 300 and 400 K, which covers the normal operating range of a semiconductor chip. The required power supply voltage is 0.3 V with a typical power consumption of 150 μW. The linearity is 0.5 ohm/degree and the surface area is 300 nm × 20 nm. Because the working principle of the device is based on integration of domain wall motion under thermal

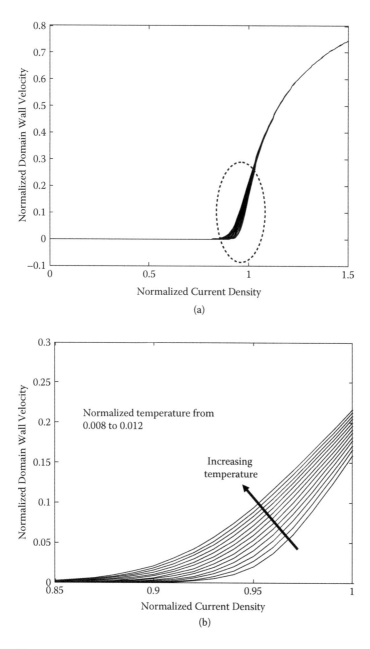

FIGURE 5.20
Average domain wall velocity as a function of normalized current density for different thermal fluctuation strengths: (a) and (b) are the same figure plotted at different normalized current density scales.

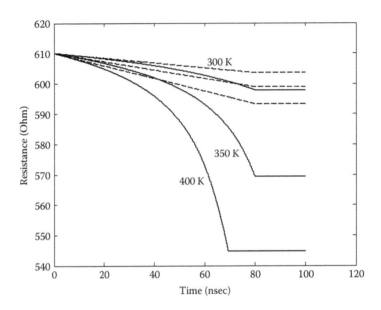

FIGURE 5.21
Spintronic memristor resistance as a function of time at different temperatures (300, 350, 400 K) for a constant magnitude voltage pulse driving.

fluctuation, the intrinsic random noise variations are averaged and suppressed. The device provides reproducible temperature readings. Compared to the available on-chip temperature sensors, a spintronic memristor requires much lower power supply voltage (which can even be used in the subthreshold voltage region) and power consumption (compared to an on-chip oscillation ring–based temperature sensor that consumes a lot of dynamic power), nanoscale feature size, low cost, and the mature integration technology with CMOS processes. Such a device is targeted for highly integrated on-chip thermal detection applications; for example, cell size < 1 μm².

References

[1] Chua, L. (1971). Memristor—The missing circuit element. *IEEE Transactions on Circuit Theory*, 18(5): 507–519.
[2] Strukov, D. B., Snider, G. S., Stewart, D. R., and Williams, R. S. (2008). The missing memristor found. *Nature*, 453: 80–83.
[3] Ho, Y., Huang, G. M., and Li, P. (2009). Nonvolatile memristor memory: Device characteristics and design implications. Paper presented at the International Conference on Computer-Aided Design, San José, CA.
[4] Choi, H., Jung, H., Lee, J., Yoon, J., Park, J., Seong, D.-J., Lee, W., Hasan, M., Jung, G.-Y., and Hwang, H. (2009). An electrically modifiable synapse array of resistive switching memory. *Nanotechnology*, 20(34): 345201.

[5] Pershin, Y. V. and Di Ventra, M. (2009). Experimental demonstration of associative memory with memristive neural networks. Available at: http://hdl.handle.net/10101/npre.2009.3258.1 (accessed February 3, 2011).

[6] Lehtonen, E. and Laiho, M. (2009). Stateful implication logic with memristors. Paper presented at the International Symposium on Nanoscale Architectures, San Francisco, CA.

[7] Chua, L. O. and Kang, S. M. (1976). Memristive devices and systems. *Proceedings of the IEEE*, 64: 209–223.

[8] Wang, J., Sun, B., Gao, F., and Greenham, N. C. (2010). Memristive devices based on solution-processed ZnO nanocrystals. *Physica Status Solidi A*, 207(2): 484–487.

[9] Wang, X., Chen, Y., Xi, H., Li, H., and Dimitrov, D. (2009). Spintronic memristor through spin-torque–induced magnetization motion. *IEEE Electron Device Letters*, 30: 294–297.

[10] Pershin, Y. V. and Ventra, M. D. (2008). Spin memristive systems: Spin memory effects in semiconductor spintronics. *Physical Review B*, 78: 113309.

[11] Erokhin, V. and Fontana, M. P. (2008). Electrochemically controlled polymeric device: A memristor (and more) found two years ago. Available at: http://arxiv.org/abs/arXiv:0807.0333 (accessed February 3, 2011).

[12] Kriegerand, J. H. and Spitzer, S. M. (2004). Non-traditional, non-volatile memory based on switching and retention phenomena in polymeric thin films. Paper presented at the Non-Volatile Memory Technology Symposium, Orlando, FL.

[13] Huai, Y. (2008). Spin-transfer torque MRAM (STT-MRAM): Challenges and prospects. *APPS Bulletin*, 18: 33.

[14] Krzysteczko, P., Reiss, G., and Thomas, A. (2009). Memristive switching of MgO based magnetic tunnel junctions. *Applied Physics Letters*, 95(11): 112508–112508-3.

[15] Gullapalli, K. K., Tsao, A. J., and Neikirk, D. P. (1993). Multiple self-consistent solutions at zero bias and multiple conduction curves in quantum tunneling diodes incorporating N^--N^+-N^- spacer layers. *Applied Physics Letters*, 62: 2971–2973.

[16] Pino, R. E., Bohl, J. W., McDonald, N., Wysocki, B., Rozwood, P., Campbell, K. A., Oblea, A., and Timilsina, A. (2010). Compact method for modeling and simulation of memristor devices: Ion conductor chalcogenide-based memristor devices. Paper presented at the International Symposium on Nanoscale Architectures (NANOARCH), Anaheim, CA.

[17] Akimoto, H., Kanai, H., Uehara, Y., Ishizuka, T., Kameyama, S. (2005). Analysis of thermal magnetic noise in spin-valve GMR heads by using micromagnetic simulation. *Journal of Applied Physics*, 97(10): 10N705–10N705-3.

[18] Bazaliy, Y. B., Jones, B. A., and Zhang, S.-C. (1998). Modification of the Landau-Lifshitz equation in the presence of a spin-polarized current in colossal- and giant-magnetoresistive materials. *Physical Review B*, 57(6): R3213–R3216.

[19] Thiavill, A., Nakatani, Y., Milta, J., and Suzuki, Y. (2005). Micromagnetic understanding of current-driven domain wall motion in patterned nanowires. *Europhysics Letters*, 69(6): 990.

[20] Zhang, S. and Li, Z. (2004). Roles of nonequilibrium conduction electrons on the magnetization dynamics of ferromagnets. *Physical Review Letters*, 93(12): 127204.

[21] Li, Z. and Zhang, S. (2004). Domain-wall dynamics driven by adiabatic spin-transfer torques. *Physical Review B*, 70(2): 024417.

[22] Liu, X., Liu, X.-J., and Ge, M.-L. (2005). Dynamics of domain wall in a biaxial ferromagnet interacting with a spin-polarized current. *Physical Review B*, 71(22): 224419.

[23] Tatara, G. and Kohno, H. (2004). Theory of current-driven domain wall motion: Spin transfer versus momentum transfer. *Physical Review Letters*, 92(8): 086601.

[24] Ho, Y., Huang, G. M., and Li, P. (2009). Nonvolatile memristor memory: Device characteristics and design implications. Paper presented at the International Conference on Computer-Aided Design, San José, CA.

[25] Tour, J. M. and He, T. (2008). Electronics: The fourth element. *Nature*, 453: 42–43.

[26] Niu, D., Chen, Y., and Xie, Y. (2010). Low-power dual-element memristor-based memory design. Paper presented at the International Symposium on Low Power Electronics and Design, Austin, TX.

[27] Wang, X., Chen, Y., Gu, Y., and Li, H. (2010). Spintronic memristor temperature sensor. *IEEE Electron Device Letters*, 31(1): 20–22.

[28] Brown, W. F. (1963). Thermal fluctuations of a single-domain particle, *Physical Review*, 130: 1677–1686.

[29] Slonczeski, J. C. (1996). Current driven excitation of magnetic multilayers, *Journal of Magnetism and Magnetic Materials*, 159: L1–L7.

[30] Berger, L. (1996). Emission of spin waves by magnetic multilayer traversed by a current. *Physical Review B*, 54(13): 9353–9358.

[31] Duine, R. A., Nunez, A. S., and MacDonald, A. H. (2007). Thermally assisted current-driven domain-wall motion. *Physical Review Letters*, 98: 056601.

[32] Williams, R. S. (2008). How we found the missing memristor. *IEEE Spectrum*, 45: 28–35.

[33] Hu, M., Li, H., Chen, Y., and Wang, X. (2011). Spintronic memristor: Compact model and statistical analysis. *Journal of Low Power Electronics*, 7: 234–244.

6

The Future of Nonvolatile Memory

Motivation for research on new types of nonvolatile memory stems from concerns regarding NAND flash memory scaling. The latest technology node used in NAND flash manufacturing has entered sub-28-nm range, which is far more advanced than 20 years ago when emerging nonvolatile memory technologies were first investigated. On the one hand, some physical limitations have been significantly improved by the technology advances; for example, the reliability degradation due to the gate oxide thickness reduction has been improved by the application of high-K materials. On the other hand, the intrinsic limitations of all nanoscale devices—for example, process variations, etc.—have already generated visible impacts on all emerging nonvolatile memory technologies.

Two types of competition exist with regard to technology scaling of emerging memory technology: (1) competition with NAND flash and (2) competition with other emerging memory technologies. In this chapter, we will summarize the latest developments in emerging nonvolatile memory technologies and offer a best guess as to the future of the emerging memory technologies discussed in this book.

6.1 Developments in Future Nonvolatile Memory Technologies

Almost every major semiconductor company in the world has established research programs on the emerging memory technologies.

6.1.1 Review of Current Efforts in the World

As the only candidate for embedded applications, research on spin-transfer torque random access memory (STT-RAM) has gained a lot of support globally. The majority of STT-RAM research is occurring in the United States and Japan, and companies involved include IBM [1], Intel [2], Seagate [3], Freescale (now Everspin) [4], Qualcomm [5], Grandis [6], Hitachi [7], Sony [8], Toshiba [9], Fujitsu [10], NEC [11], etc. The capacity of prototyping chips has grown to to 64 Mb [12]. When the technology scales, the write current of a magnetic tunneling junction (MTJ) reduces quadratically (or linearly with device area), as shown in Figure 6.1 [13]. This trend offers a good scalability of STT-RAM in terms of write power consumption and density. More important, the reduced write current can be provided by an N-channel metal-oxide

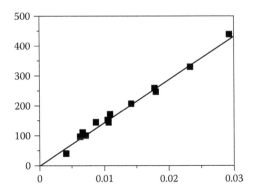

FIGURE 6.1
Write current reduction when MTJ area is decreased [13].

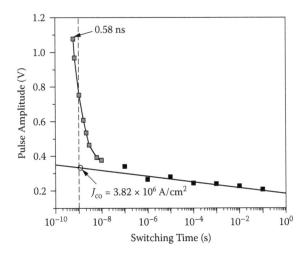

FIGURE 6.2
Switching performance of in-plane MTJ under 1 ns [14].

semiconductor field-effect transistor (NMOS) with minimal size. A $6F^2$ STT-RAM cell area can be achieved.

The latest experiments have shown that even for a traditional MTJ structure (known as in-plane MTJ), the switching speed can be shorter than 1 ns or less [14], as shown in Figure 6.2. Zhao et al. [13] predicted two promising techniques to further improve the switching speed (or reduce the switching current), including a (1) perpendicular MTJ and (2) dual-barrier MTJ. For example, a perpendicular MTJ is expected to offer a fivefold improvement in write current in a sub-nanosecond switching region [13].

However, we shall point out that continuous scaling of the write current does not necessarily benefit STT-RAM design. Following a decrease of the

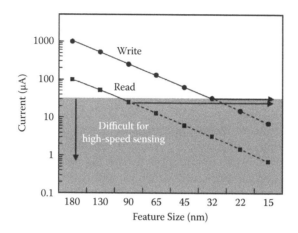

FIGURE 6.3
Feature size dependence of write current and read current [15].

write current, the read current must be reduced accordingly to avoid disturbance of the memory bits. However, the scaled read current leads to a smaller
sense margin, which may not be sufficient for high-speed sensing, as shown
in Figure 6.3 [15].

As the technology scales, the process variation–induced MTJ resistance variations will become more significant. As a result, the sense margin will continue to shrink and incur longer read latency. This trend will be the primary
technical difficulty in STT-RAM designs under the scaled technology nodes.

Up to the completion of this book, commercial products with phase change
memory (PCM) up to 128 Mb have become available through micrometer
technology. Higher capacity chips have also been presented, including 256
Mb [16] and 512 Mb [17]. In particular, a memory cell area as low as $5.8F^2$ was
achieved by K.-J. Lee et al. [16], which is at the same level of dynamic RAM
(DRAM) technology.

However, there are two major issues that make PCM less competitive compared to STT-RAM and resistive RAM (R-RAM) in the embedded applications: (1) the slow access latency and (2) the low endurance cycles. In both
K.-J. Lee et al. [16] and De Sandre et al. [17], the read latencies were longer
than 55 ns and the write latencies were as long as 600 ns (reset operation).
The endurance cycle was only 10^5. In the latest PCM demonstration, the read
latency and endurance performance were improved to 12 ns (4 Mb memory)
and 10^6 by paying significant area overhead ($36F^2$ cell area) [18].

Very recently, a 64 Mb R-RAM chip was first demonstrated [19]. A higher
capacity chip—that is, 64 Gb—is also under development [20]. HP Labs will
also unveil prototype chips of nonvolatile memory based on memristor
technology using crossbar arrays [21]. However, the biggest issue in R-RAM
development is the uncertainty in the material options: too many materials
can be chosen but no materials can meet all of the expected requirements.

6.2 Future Research Trends

Although many unique characteristics have been demonstrated by the latest research on the next generation of nonvolatile memories, there are still some technical obstacles that need to be overcome before they enter the mass production stage.

As the most mature technology, the future research on PCM may focus on the improvement of memory product specifications; that is, access and endurance performance. In addition to the development of material and devices, more architecture-level works are emerging. For example, in Yang et al. [22], the authors presented many architectural designs that can make access to a PCM-based memory uniform and enhance its lifetime, as shown in Figure 6.4.

As discussed, the reduction of write current and the increase in MTJ variations point toward a mixed future for STT-RAM development. The targeted applications of STT-RAM require a fast read access, which is very hard to realize under the small sense margin. In addition, the high write access error, which is mainly incurred by process variations and thermal fluctuations, can be another limiting factor in STT-RAM design. Therefore, future research on STT-RAM may focus on enhancement of access performance and error reductions.

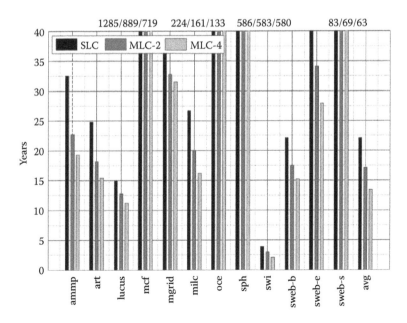

FIGURE 6.4
PCM lifetime after (1) eliminating redundant bit writes, (2) row shifting, and (3) segment swapping [22].

Research on R-RAM has achieved significant accomplishments. More than 10^{10} endurance cycles have been demonstrated with a programming current less than 5 mA. As the device yield approaches 100%, the focus on R-RAM development will shift to chip-level integration, which requires not only the design of a peripheral circuit but improvement of device uniformity.

6.2.1 Who Will Be the Winner?

There are many examples of unsuccessful predictions regarding future of technologies. The reason is very simple: prediction is always based on extrapolation of the current trend, which is highly unreliable due to many uncertainties in technology development. As a summary of this book, however, we will try to answer this tough question based on our general understanding of the memory technologies introduced.

6.2.2 Semiconductor Memory Roadmap

Table 6.1 shows relevant predictions regarding STT-RAM, PCM, and R-RAM proposed by the International Technology Roadmap for Semiconductors (ITRS). It shows that all of the merging memories will target the 5-6F^2 cell area. However, PCM has less attractive properties in terms of access performance, endurance, and operating voltage; in addition, the write power of R-RAM is not yet known. If these parameters cannot be significantly improved in the near future, PCM may be replaced by R-RAM (or even STT-RAM) for massive data storage applications. Although the write energy of R-RAM is unknown, we believe that it will be significantly lower than either PCM or STT-RAM due to the relatively low operating voltage and high resistance range.

6.2.3 Forecast

To a certain degree, the future of emerging nonvolatile memory technologies relies mainly on investment of resources. The physical fundamentals of all

TABLE 6.1

Predictions of Emerging Memory Technologies

	PCM	STT-RAM	R-RAM
Cell area	4.7F^2	6F^2	5F^2
Read time	<60 ns	<10 ns	<10 ns
W/E time	<50 ns	<2 ns	<10 ns
Retention time	>10 Years	>10 Years	>10 Years
Write cycles	>10^{15}	>10^{16}	>10^{16}
Write operating voltage (V)	<3	<1	<0.5
Read operating voltage (V)	<3	<0.5	<0.5
Write energy (J/bit)	<10^{-13}	<10^{-13}	Unknown

of the memory technologies introduced have been well studied and their development has entered the engineering demonstration stage (except for PCM). However, the major memory companies face hard decisions between continuing to invest in existing technology—for example, NAND flash—and investing in new memory technologies. Unless they see overwhelming benefits from the new memory technologies, they will not be able to change the current investment allocations.

We believe that the adoption of emerging memory technologies will be driven by the benefits demonstrated by progress in recent research and the demand for unique memory properties; for example, random and fast access of nonvolatile memories. In fact, we have already seen such a trend in recent years, and we are optimistic.

References

[1] Gallagher, W. J. S. and Parkin, S. P. (2006). Development of the magnetic tunnel junction MRAM at IBM: From first junctions to a 16-Mb MRAM demonstrator chip. *IBM Journal of Research & Development,* 50(1): 5–23.

[2] Raychowdhury, A., Somasekhar, D., Karnik, T., and De, V. (2009). Design space and scalability exploration of 1T-1STT MTJ memory arrays in the presence of variability and disturbances. Paper presented at the International Electron Device Meeting, Baltimore, MD.

[3] Chen, Y., Wang, X., Li, H., Liu, H., and Dimitar, D. (2008). Design margin exploration of STT-MRAM. Paper presented at the International Symposium on Quality Electronic Design, San José, CA.

[4] Andre, T. W., Nahas, J. J., Subramanian, C. K., Garni, B. J., Lin, H. S., Omair, A., and Martino, W. L. (2005). 4-Mb 0.18-mm 1T1MTJ toggle MRAM with balanced three input sensing scheme and locally mirrored unidirectional write drivers. *IEEE Journal of Solid-State Circuits,* 40(1): 301–309.

[5] Lin, C., Kang, S., Wang, Y., Lee, K., Zhu, X., Chen, W., Li, X., Hsu, W., Kao, Y., Liu, M., Lin, Y., Nowak, M., Yu, N., and Tran, L. (2009). 45 nm low power CMOS logic compatible embedded STT MRAM utilizing a reverse-connection 1T/1MTJ cell. Paper presented at the IEEE International Electron Devices Meeting, Baltimore, MD.

[6] Diao, Z., Li, Z., Wang, S., Ding, Y., Panchula, A., Chen, E., Wang, L., and Huai, Y. (2007). Spin-transfer torque switching in magnetic tunnel junctions and spin-transfer torque random access memory. *Journal of Physics: Condensed Matter,* 19: 165209.

[7] Kawahara, T., Takemura, R., Miura, K., Hayakawa, J., Ikeda, S., Lee, Y., Sasaki, R., Goto, Y., Ito, K., Meguro, I., Matsukura, F., Takahashi, H., Matsuoka, H., and Ohno, H. (2007). 2 Mb spin-transfer torque RAM (SPRAM) with bit-by-bit bidirectional current write and parallelizing-direction current read. Paper presented at the IEEE International Solid-State Circuits Conference, San Francisco, CA.

[8] Hosomi, M., Yamagishi, H., Yamamoto, T., Bessho, K., Higo, Y., Yamane, K., Yamada, H., Shoji, M., Hachino, H., Fukumoto, C., Nagao, H., and Kano, H. (2005). A novel nonvolatile memory with spin torque transfer magnetization switching: Spin-RAM. Paper presented at the IEEE International Electron Devices Meeting, Washington, DC.

[9] Tsuchida, K., Inaba, T., Fujita, K., Ueda, Y., Shimizu, T., Asao, Y., Kajiyama, T., Iwayama, M., Sugiura, K., Ikegawa, S., Kishi, T., Kai, T., Amano, M., Shimomura, N., Yoda, H., and Watanabe, Y. (2010). A 64 Mb MRAM with clamped-reference and adequate-reference schemes. Paper presented at the IEEE International Solid-State Circuits Conference, San Francisco, CA.

[10] Halupka, D., Huda, S., Song, W., Sheikholeslami, A., Tsunoda, K., Yoshida, C., and Aoki, M. (2010). Negative resistance read and write schemes for STT-MRAM in 0.13 mm CMOS. Paper presented at the IEEE International Solid-State Circuits Conference, San Francisco, CA.

[11] Nebashi, R., Sakimura, N., Honjo, H., Saito, S., Ito, Y., Miura, S., Kato, Y., Mori, K., Ozaki, Y., Kobayashi, Y., Ohshima, N., Kinoshita, K., Suzuki, T., Nagahara, K., Ishiwata, N., Suemitsu, K., Fukami, S., Hada, H., Sugibayashi, T., and Kasai, N. (2009). A 90 nm 12 ns 32 Mb 2T1MTJ MRAM. Paper presented at the IEEE International Solid-State Circuits Conference, San Francisco, CA.

[12] Driskill-Smith, A., Watts, S., Apalkov, D., Druist, D., Tang, X., Diao, Z., Luo, X., Ong, A., Nikitin, V., and Chen, E. (2010). Non-volatile spin-transfer torque RAM (STT-RAM): An analysis of chip data, thermal stability and scalability. Paper presented at the Memory Workshop (IMW), 2010 IEEE International, Monterey, CA.

[13] Zhao, H., Lyle, A., Zhang, Y., Amiri, P. K., Rowlands, G., Zeng, Z., Katine, J., Jiang, H., Galatsis, K., Wang, K. L., Krivorotov, I. N., and Wang, J.-P. (2011). Low writing energy and sub nanosecond spin torque transfer switching of in-plane magnetic tunnel junction for spin torque transfer random access memory. *Journal of Applied Physics*, 109: 07C720.

[14] Ono, K., Kawahara, T., Takemura, R., Miura, K., Yamamoto, H., Yamanouchi, M., Hayakawa, J., Ito, K., Takahashi, H., Ikeda, S., Hasegawa, H., Matsuoka, H., and Ohno, H. (2009). A disturbance-free read scheme and a compact stochastic-spin-dynamics-based MTJ circuit model for Gb-scale SPRAM. Paper presented at the 2009 IEEE International Electron Devices Meeting (IEDM), 7–9, Baltimore, MD.

[15] Kang, S., Cho, W., Cho, B.-H., Lee, K.-J., Lee, C.-S., Oh, H.-R., Choi, B.-G., Wang, Q., Kim, H.-J., Park, M.-H., Ro, Y.-H., Kim, S., Kim, D.-E., Cho, K.-S., Ha, C.-D., Kim, Y., Kim, K.-S., Hwang, C.-R., Kwak, C.-K., Byun, H.-G., and Shin, Y. S. (2006). A 0.1 mm 1.8 V 256 Mb 66 MHz synchronous burst PRAM. Paper presented at the IEEE International Solid-State Circuits Conference, San Francisco, CA.

[16] Lee, K.-J., Cho, B.-H., Cho, W.-Y., Kang, S., Choi, B.-G., Oh, H.-R., Lee, C.-S., Kim, H.-J., Park, J.-M., Wang, Q., Park, M.-H., Ro, Y.-H., Choi, J.-Y., Kim, K.-S., Kim, Y.-R., Shin, I.-C., Lim, K.-W., Cho, H.-K., Choi, C.-H., Chung, W.-R., Kim, D.-E., Yu, K.-S., Jeong, G.-T., Jeong, H.-S., Kwak, C.-K., Kim, C.-H., and Kim, K. (2007). A 90 nm 1.8 V 512 Mb diode-switch PRAM with 266 MB/s read throughput. Paper presented at the IEEE International Solid-State Circuits Conference, San Francisco, CA.

[17] De Sandre, G., Bettini, L., Pirola, A., Marmonier, L., Pasotti, M., Borghi, M., Mattavelli, P., Zuliani, P., Scotti, L., Mastracchio, G., Bedeschi, F., Gastaldi, R., and Bez, R. (2010). A 90 nm 4 Mb embedded phase change memory with 1.2 V 12 ns read access time and 1 MB/s write throughput. Paper presented at the IEEE International Solid-State Circuits Conference, San Francisco, CA.

[18] Chevallier, C., Siau, C. H., Lim, S., Namala, S., Matsuoka, M., Bateman, B., and Rinerson, D. (2010). A 0.13 mm 64 Mb multi-layered conductive metal-oxide memory. Paper presented at the IEEE International Solid-State Circuits Conference, San Francisco, CA.

[19] LaPedus, M. (2009). Unity rolls "storage class" memory technology. Available at: http://www.eetimes.com/electronics-news/4083161/Unity-rolls-storage-class-memory (accessed April 1, 2011).

[20] Johnson, R. C. (2008). Memristors ready for prime time. Available at: http://http://www.eetimes.com/electronics-news/4077811/Memristors-ready-for-prime-time (accessed April 1, 2011).

[21] Lee, B. C., Zhou, P., Yang, J., Zhang, Y., Zhao, B., Ipek, E., Mutlu, O., and Burger, D. (2010). Phase-change technology and the future of main memory. *IEEE Micro Magazine*, 30(1): 143.

[22] Yang, J. J., Zhang, M., Strachan, J. P., Borghetti, J., Pickett, M. D., Miao, F., Xia, Q., Ohlberg, D. A. A., Nickel, J. H., Ribeiro, G. M., and Williams, R. S. (2011). Recent progress on oxide based memristive devices in HP. Paper presented at the 2011 Non-volatile Memories Workshop, San Diego, CA.

Index

For Product Safety Concerns and Information please contact our EU
representative GPSR@taylorandfrancis.com
Taylor & Francis Verlag GmbH, Kaufingerstraße 24, 80331 München, Germany

www.ingramcontent.com/pod-product-compliance
Ingram Content Group UK Ltd.
Pitfield, Milton Keynes, MK11 3LW, UK
UKHW020952180425
457613UK00019B/651